BRITISH FRESHWATER MEGALOPTERA AND NEUROPTERA:

A KEY WITH ECOLOGICAL NOTES

by

J. M. ELLIOTT

Freshwater Biological Association and Institute of Freshwater Ecology

Illustrated by
D. E. KIMMINS, C. JOAN WORTHINGTON
and J. M. ELLIOTT

FRESHWATER BIOLOGICAL ASSOCIATION
SCIENTIFIC PUBLICATION No. 54

1996

Series Editor: J. M. ELLIOTT

The National Rivers Authority (NRA) welcomes the opportunity to part-sponsor the publication of *British Freshwater Megaloptera and Neuroptera: A Key with Ecological Notes*, the latest in the FBA's series of Scientific Publications which continue to provide essential authoritative information needed to underpin the management of freshwater habitats. The publication is particularly relevant to the NRA because the alder-flies, snake-flies, sponge-flies and lacewings include a number of taxa with aquatic larvae that are used to assess water quality by biological means.

The key and associated notes represent an invaluable source reference for identification, distribution and ecology. High quality information is required to maximise the effectiveness of measures to assess and monitor the state of rivers as part of the Authority's aim to achieve a continuing improvement in the water environment in England and Wales. Moreover, accurate identification of rarer macroinvertebrates is essential with respect to the Authority's duty to further conservation.

The NRA strongly recommends this FBA publication to practitioners and enthusiasts alike, but the NRA accepts no responsibility whatsoever for any errors or mis-statements contained in this publication. Any queries or comments should be directed to the FBA.

Published by the Freshwater Biological Association,
The Ferry House, Far Sawrey, Ambleside, Cumbria LA22 0LP

ISBN 0 900386 56 8

ISSN 0367-1887

PREFACE

The original publication on aquatic Megaloptera and Neuroptera was written by D. E. Kimmins and published by the Association in 1944, with a second edition in 1962. A new edition by the present author was produced in 1977 and this included additional illustrations and an expanded text. As adults of species with terrestrial larvae may be confused with adults of species with aquatic larvae, a key to all families was included in the new edition and some of the keys to species were revised. Since the publication of the edition of 1977, a third species of alder-fly has been added to the British list and some advance has been made in identifying larvae of sponge-flies. These changes are incorporated in this new edition. The section on life cycles and ecology has been expanded and is now a more complete review of the literature. The list of references includes all the important publications and should therefore serve as a guide to the literature.

We are most grateful to the National Rivers Authority for sponsoring the publication of this new key which will be of value to those who use these species as indicators of water quality, and to those with an interest in the conservation of rare species such as the sponge-flies and the largest British lacewing, *Osmylus fulvicephalus*.

The Ferry House
September 1995

J. M. Elliott
Series Editor

CONTENTS

INTRODUCTION

The Megaloptera and Neuroptera (or Planipennia) are usually regarded as two closely related orders of the Insecta and some specialists combine them into one order of Neuroptera. Both orders belong to the Endopterygota (or Holometabola), i.e. those insects with a complete metamorphosis internally in the larval and pupal stages, with a larva that differs from the adult in its structure and habits, and with a life cycle divided into four definite stages of egg, larva, pupa and imago. The name Neuroptera (Greek *neuron*, nerve; *pteron*, wing) refers to the nerve-like network of veins in the wings. Linnaeus originally used this name for all insects with delicate, net-veined wings and the Linnaean order included the exopterygote Ephemeroptera and Odonata. This miscellany of veined-winged insects was gradually split into several orders and now the order Neuroptera includes a relatively small number of species.

The Megaloptera and Neuroptera are among the most primitive of the endopterygote insects, being close to the ancestral stock of the Panorpoid complex which also includes the Mecoptera, Lepidoptera, Trichoptera, Siphonaptera and Diptera. Their characteristic wings are very similar to those of Palaeozoic fossil insects and the first fossil records of the two orders are from the Lower Permian of Kansas. Megalopteran wings, resembling those of alder-flies, have been described from the Upper Permian of European Russia and a young sialid-like larva was found in amber from the Upper Cretaceous of the Taimir Peninsula (Wootton 1988). A sub-fossil head of an alder-fly larva was recorded from the interglacial strata in the province of Jämtland, Sweden (Tjeder 1968). About 4500 species are known at present to belong to the two orders of insects and sixty species have been recorded in Britain..

GENERAL CHARACTERS

Adults of both orders are usually soft-bodied with long slender antennae, with no abdominal cerci, with their tarsi having five segments (the tarsus is the distal part of the leg and usually ends in a pair of claws), and with their mouthparts adapted for biting. They have two pairs of very similar membraneous wings which are held at rest like a peaked roof or tent over the abdomen. The hindwings are only slightly smaller than the forewings and are of similar shape. Although the wing venation is generally regarded as primitive, there are usually many accessory veins and numerous costal veinlets giving a ladder-like effect along the anterior edge of the wing. The radial sector in the forewing (third longitudinal vein behind the anterior edge of the wing) emits two or more branches. Adult Neuroptera are usually crepuscular but adult Megaloptera fly readily during warm sunny weather. Species with aquatic larvae do not lay their eggs in water but deposit them on objects overhanging the water.

Pupae of both orders are exarate, i.e. the appendages are not adherent to the rest of the body and can be used in locomotion. Species with aquatic larvae pupate on land but always close to water.

Larvae of both orders have a well-sclerotized head with obvious forwardly-directed mouthparts and large "jaws" (mandibles). Their thoracic legs are usually long and there are no abdominal prolegs. The whole body or at least the abdomen is more or less elongately fusiform in shape. Aquatic larvae have seven pairs of lateral or ventral abdominal gills. All larvae are active predaceous insects with biting mouthparts in the Megaloptera and sucking mouthparts in the Neuroptera.

CLASSIFICATION

The British species of Megaloptera fall naturally into the suborders Raphidioidea and Sialoidea, and each suborder has only one family, namely the Raphidiidae with four species in the genus *Raphidia*, and the Sialidae with three species in the genus *Sialis*. Larvae of *Raphidia* are terrestrial and the adults are known as snake-flies because of their narrow elongate prothorax which forms a kind of "neck" on which the head is held like that of a snake ready to strike. Larvae of *Sialis* are aquatic and the adults are known as alder-flies, no doubt because they are often found on alders that overhang the water.

The British species of Neuroptera fall into five families, namely the Coniopterygidae with seven species in five genera, the Osmylidae with only one species, the Sisyridae with three species in one genus, the Hemerobiidae with twenty-nine species in eight genera and the Chrysopidae with fourteen species in two genera. Adults of the two largest families are known as the brown (Hemerobiidae) and green (Chrysopidae) lacewings, and their larvae are always terrestrial. Larvae of the Coniopterygidae are also terrestrial and the minute adults are the smallest and most aberrant of the Neuroptera with their opaque white, waxy wings which have a greatly reduced venation. Larvae of the Sisyridae are aquatic and the adults are sometimes known as sponge-flies because the larvae are usually found on freshwater sponges on which they feed. The Osmylidae are closely allied to the Sisyridae and larvae of the single British species are semi-aquatic, living under stones or in moss at the edges of streams.

Adults of the British species of Megaloptera and Neuroptera were described by McLachlan (1868) and Killington (1929), and keys to adults are included in the handbook of Fraser (1959). Killington (1936, 1937) has also written a monograph on the British Neuroptera. Barnard (1978) provides a check-list of the British Neuroptera with some taxonomic notes. Only three of the seven species of Megaloptera and four of the fifty-four species of Neuroptera are aquatic or semi-aquatic in the larval stage. These seven species are given in the following checklist.

CHECKLIST

MEGALOPTERA
- Family SIALIDAE
  - Genus *Sialis* Latreille, 1803
    - *S. lutaria* (Linnaeus, 1758)
      (=*flavilatera* Linnaeus, 1758)
    - *S. fuliginosa* Pictet, F.-J., 1836
    - *S. nigripes* Pictet, A.-E., 1865

NEUROPTERA
- Family OSMYLIDAE
  - Genus *Osmylus* Latreille, 1802
    - *O. fulvicephalus* (Scopoli, 1763)
      (= *chrysops* Linnaeus, 1758)
      (= *maculatus* Fabricius, 1787)
      (= *laurifoliaeformis* Razoumowsky, 1789)

- Family SISYRIDAE
  - Genus *Sisyra* Burmeister, 1839
    - *S. fuscata* (Fabricius, 1793)
      (= *confinis* Stephens, 1836)
      (= *morio* Burmeister, 1839)
      (= *nigripennis* Wesmael, 1841)
      (= *spongillae* Westwood, 1842)
      (=*fumatus* Motschulsky, 1853)
    - *S. dalii* McLachlan, 1866
    - *S. terminalis* Curtis, 1854

COLLECTION AND PRESERVATION

Adults are most readily collected by beating bushes and trees, or sweeping low herbage near ponds, lakes and streams. *Osmylus* is sometimes found on the underside of low bridges over woodland streams. Some kind of bottom sampler is required to collect larvae of *Sialis* from the mud of ponds, lakes and streams, and a guide to such samplers is provided by Elliott, Tullett & Elliott (1993). Larvae occasionally may be found under large stones in streams. *Sisyra* larvae are found in freshwater sponges of lakes, canals and streams. *Osmylus* larvae live in the moss bordering woodland streams.

If a pinned collection is desired, adults should be brought back alive and killed with ether or chloroform. Stainless steel pins and flat setting boards should be used, the specimen being set in the same way as Lepidoptera, the wings being held in place by strips of paper or cellophane during drying. They should be set as soon as possible after killing, since *Sisyra*, in particular, dries very rapidly and becomes brittle.

Adults, pupae and larvae may also be preserved in fluid, e.g. 70% alcohol, dilute formaldehyde (one part of 40% formaldehyde to nineteen parts of water), or a mixture of alcohol and dilute formaldehyde. A suitable fixative fluid is Pämpel's mixture (four parts of glacial acetic acid, thirty parts of distilled water, six parts of 40% formaldehyde and fifteen parts of 95% alcohol). An excellent killing fluid and preservative for material suitable for dissection is KAAD (one part of kerosene, ten parts of 95% alcohol, two parts of glacial acetic acid, one part of dioxan).

KEY TO ADULTS

The general characters given in the Introduction separate the Megaloptera and all families of Neuroptera, except the Coniopterygidae, from all other orders of insects. The most useful character is the two pairs of membraneous wings covered with a delicate network of veins, including many cross-veins. Some species of Lepidoptera, Trichoptera, Ephemeroptera, Plecoptera, Odonata and Mecoptera have similar wings but can be distinguished from the Megaloptera and Neuroptera by the following characters. In the Lepidoptera (butterflies and moths), the wings are covered with minute scales and there is usually a coiled proboscis (tongue) on the underside of the head. In the Trichoptera (caddis-flies), the wings are usually densely covered with hairs, the hindwing is usually broader than the forewing and there are relatively few cross-veins. In the Ephemeroptera (mayflies), the hindwings, when present, are always considerably smaller than the forewings and there are two or three abdominal cerci. In the Odonata (dragonflies and damsel-flies), the antennae are always very short. In the Plecoptera (stoneflies), abdominal cerci are clearly seen in most species, the hindwings are usually considerably larger than the forewings and all tarsi have three segments. The Mecoptera (scorpion-flies) are closely related to the Megaloptera and Neuroptera, but are easily distinguished from them by the shape of the head which is extended downwards to form a beak or rostrum.

As the adults of species with terrestrial larvae are often found near water and may be confused with adults of species with aquatic larvae, a general key to families is given before the key to species of the three families with aquatic or semi-aquatic larvae. The Coniopterygidae are not included in the key to families because they differ considerably from other species of Neuroptera, being minute and covered with white, waxy powder. Their wing venation is greatly reduced with few cross-veins and usually with the hindwings much smaller than the forewings.

KEY TO FAMILIES

1 3rd or 4th segment of tarsus is bilobed (Fig. 1A). Branches of the wing-veins are not frequently bifurcated at the wing margin (Fig. 2A, B)— MEGALOPTERA, **2**

— All tarsal segments are similar and narrow (Fig. 1B). Branches of the wing-veins are usually conspicuously bifurcated at the margins of the wings (Fig. 2C, D, E, F)— NEUROPTERA, **3**

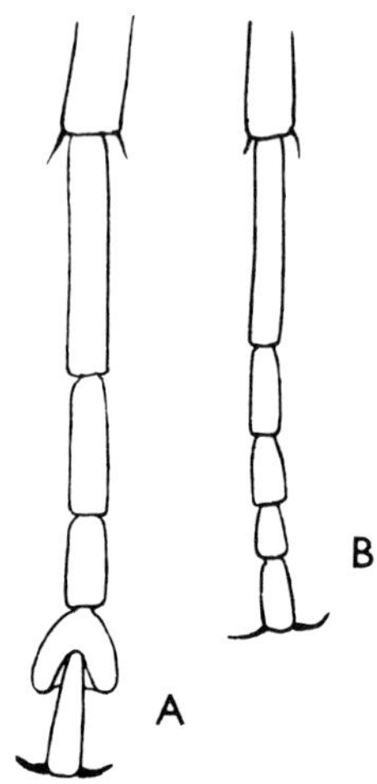

Fig. 1. Tarsi of: A, *Sialis lutaria*; B, *Sisyra fuscata*.

2 Anterior part of thorax (prothorax) elongated to form a kind of "neck" on which the head is held like that of a snake ready to strike. Wings with a pterostigma (*Pt* in Fig. 2A) which is an opaque chitinous cell. 3rd segment of tarsus is bilobed— RAPHIDIIDAE (snake-flies)

Larvae terrestrial.

— Prothorax is short and broad. Wings without a pterostigma (Fig. 2B). 4th segment of tarsus is bilobed (Fig. 1A). Wings at rest form a broad "roof" (Fig. 3A)— SIALIDAE (alder-flies), p. 14

Larvae aquatic.

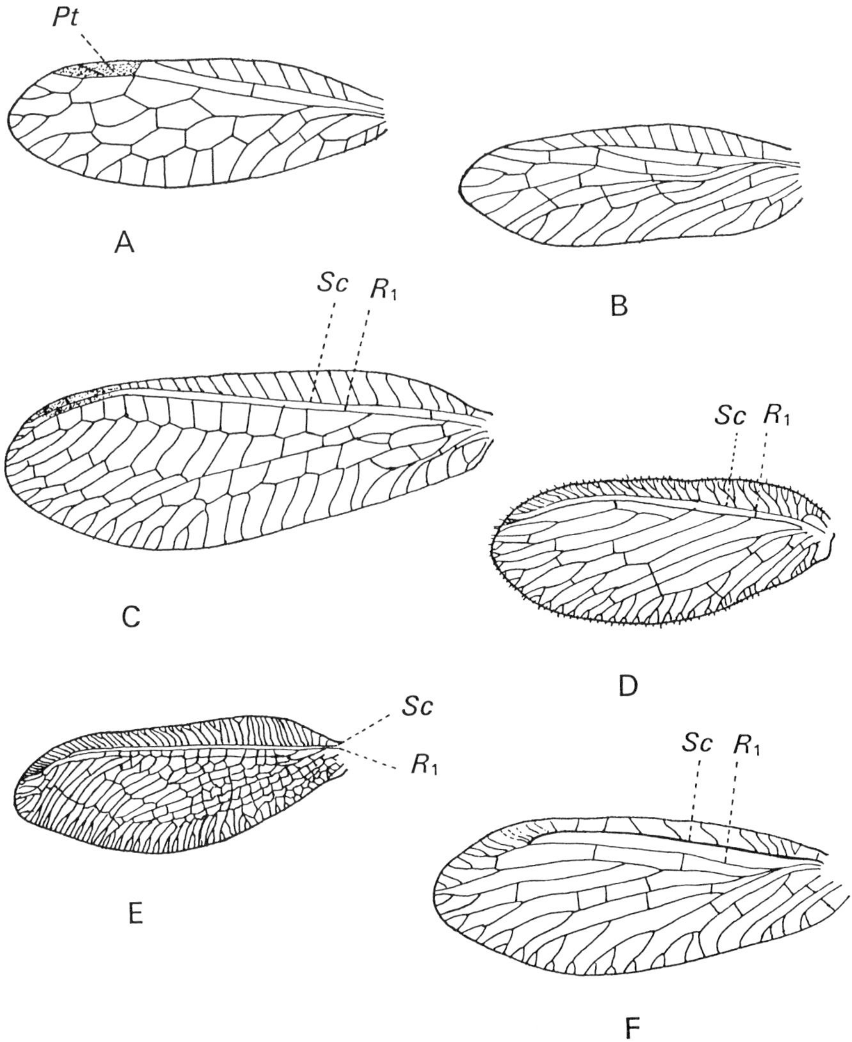

Fig. 2. Forewing of: A, Raphidiidae (length of wing 10–14 mm);B, Sialidae (8–17 mm); C, Chrysopidae (10–25 mm); D, Hemerobiidae (3–16 mm). E, Osmylidae (19–23 mm); F, Sisyridae (5–6 mm).

3 (1) Subcosta (*Sc*) and radius (R_1) do not join near the apex of the forewing (beware of cross-veins) (Fig. 2C, D) and run close together near the anterior edge of the wing— **4**

— Subcosta (*Sc*) and radius (R_1) fuse near the apex of the forewing (Fig. 2E), or the subcosta bends posteriorly to meet the radius near the apex of the forewing (Fig. 2F)— **5**

4 Large to medium-sized insects coloured yellow or bright green, with golden eyes and filiform (thread-like) antennae. There are only a few longitudinal veins in the wings (Fig. 2C) and hairs are limited to the veins, cross-veins and edge of the wing— CHRYSOPIDAE (green lacewings)

Larvae terrestrial.

— Small grey or brown insects with moniliform (like a string of beads) antennae. There are several longitudinal veins in the wings (Fig. 2D) which are covered with short fine hairs on both the veins and the membrane. Minute wart-like tubercles (trichosors), each with two or three bristles, are situated along the edge of the wing between the terminal branches of the veins— HEMEROBIIDAE (brown lacewings)

Larvae terrestrial.

5 (3) Wings are large (forewing 19–23 mm) and transparent, with numerous cross-veins. Subcosta (*Sc*) and radius (R_1) run very close together and fuse near the apex of the wing (Fig. 2E). Wing membrane has darkish brown spots or blotches (Fig. 3C)— OSMYLIDAE, p. 16

Larvae semi-aquatic.

— Wings are small (forewing 5–6 mm) with few cross-veins. Subcosta (*Sc*) bends to meet the radius (R_1) near the apex of the forewing (Figs 2F, 3B)— SISYRIDAE (sponge-flies), p. 18

Larvae aquatic.

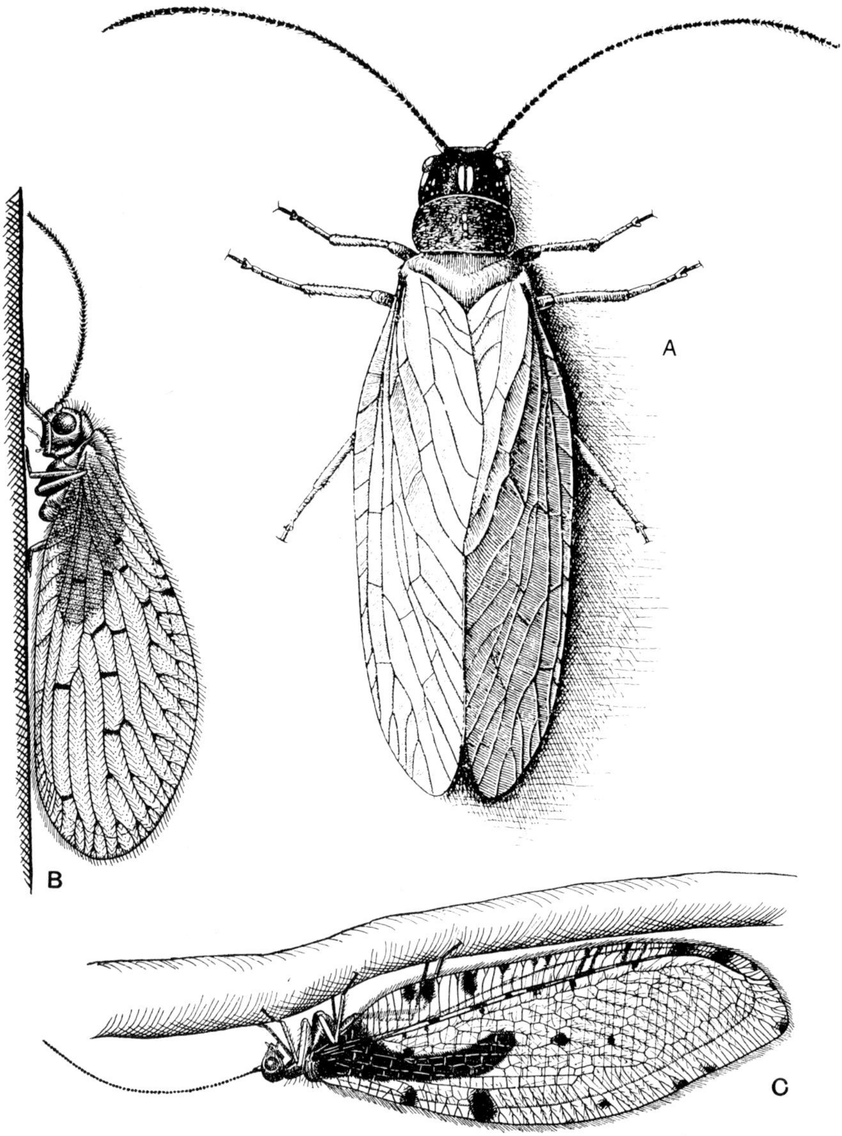

Fig. 3. Megaloptera and Neuroptera in resting positions: A, *Sialis lutaria* x 7; B, *Sisyra dalii* x 10.5; C, *Osmylus fulvicephalus* x 3.3.

KEY TO SPECIES WITH AQUATIC OR SEMI-AQUATIC LARVAE

Family SIALIDAE. Genus *Sialis* Latreille

Although the wing venation is frequently used to separate the species of *Sialis* (e.g. Fraser 1959; Kimmins 1962), it is rather variable and is therefore not a reliable character (Seitz 1940; Hoffmann 1962; Selman 1965; Vaňhara 1970). As the characters of the external genitalia are much more constant, they are more satisfactory for an unequivocal determination (Kaiser 1950; Meinander 1962; Vaňhara 1970; Barnard 1977). The sexes are easily recognised. At the end of the abdomen in the male, there are several structures adapted for mating, with the 10th tergite and 9th sternite forming the outer "pincers" (Fig. 4). Females lack similar structures, but the shape of the 8th sternite is specific (Fig. 5). Note that the terminology may differ between authors, e.g. the 10th tergite in the following key is the anal segment of Kaiser (1950), the ectoprocts of Meinander (1962), the 10th tergite of Vaňhara (1970), and the epiprocts of Barnard (1977)!

MALE GENITALIA

1 9th tergite is almost as long as 9th sternite (*9t* and *9st* in Fig. 4A). 9th sternite forms a subgenital plate which is spoon-shaped and nearly as long as broad (*9st* in Fig. 4A, B), overlapping the end of the abdomen when closed. Membraneous, tubular utriculi are well developed (*u* in Fig. 4B). Flight period late April-June, occasionally July. Adults are usually found near the banks of ponds, lakes and sluggish streams and rivers where there is an abundance of silt— **Sialis lutaria** (Linn.)

— 9th tergite is much longer than 9th sternite (*9t* and *9st* in Figs 4C, E). 9th sternite forms a short subgenital plate which is distinctly broader than long and never overlaps the end of the abdomen. Utriculi are absent— **2**

2 10th tergite (anal segment) is distinctly beak-like in profile (Fig. 4C) and forms an ovoid plate which terminates in two projections (Fig. 4D). Parameres are large and rounded with no strong spines (*p* in Fig. 4D). Flight period May-June. Adults are usually found near the banks of streams and rivers— **Sialis fuliginosa** Pict.

— 10th tergite is flat in profile (Fig. 4E) and forms a broad plate without projections (Fig. 4F). Mediuncus, ending in two short projections, may be seen projecting beyond the 10th tergite (*m* in Fig. 4F). Parameres are in two parts, the upper with distinct spines (Fig. 4F). Flight period May-June. Adults are usually found near the banks of streams and rivers— **Sialis nigripes** Pict.

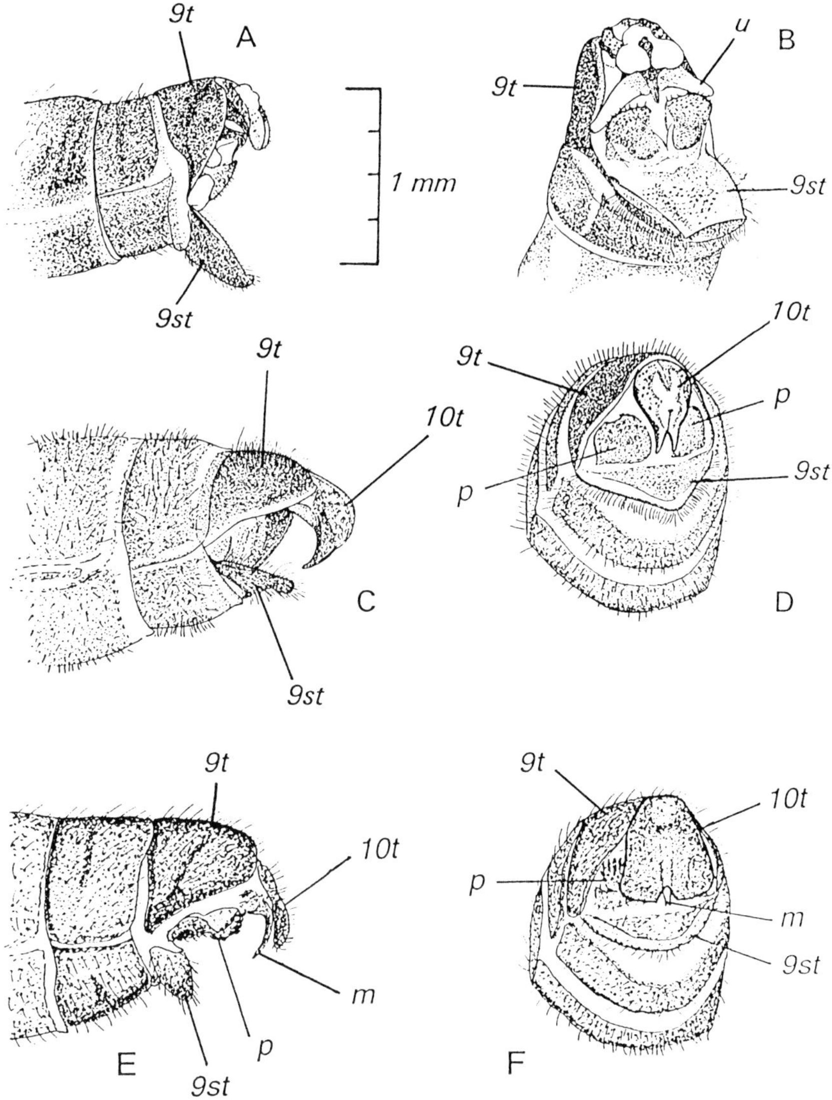

Fig. 4. External genitalia of male *Sialis*: A, *S. lutaria* in lateral view; B, *S. lutaria* from posterior; C, *S. fuliginosa* in lateral view; D, *S. fuliginosa* from posterior; E, *S. nigripes* in lateral view; F, *S. nigripes* from posterior. 9*t*, 9th tergite; 10*t*, 10th tergite; 9*st*, 9th sternite; *u*, utriculus; *p*, paramere; *m*, mediuncus.

FEMALE GENITALIA

1 Posterior edge of 7th sternite has two lateral protuberances (7*st* in Fig. 5A). 8th sternite has a depressed central area (8*st* in Fig. 5A)—
Sialis nigripes Pict.

— Posterior corners of 7th sternite are rounded (Fig. 5B) or angular (Fig. 5C). 8th sternite has a medial tooth (Fig. 5B) or narrows markedly in the mid-line (Fig. 5C)— **2**

2 8th sternite does not narrow markedly in the mid-line and has a medial tooth on the posterior margin which is slightly concave (8*st* in Fig. 5B). 7th sternite usually has rounded posterior corners—
Sialis lutaria (Linn.)

— 8th sternite narrows markedly in the mid-line and is concave towards the anterior with no medial tooth on the posterior margin (8*st* in Fig. 5C). 7th sternite usually has angular posterior corners—
Sialis fuliginosa Pict.

Family OSMYLIDAE. Genus *Osmylus* Latreille

There is only one British species (Fig. 3C). Length of forewing 19–23 mm. Flight period May-August. Adults are usually found near small woodland streams that have moss along their banks and thus provide a suitable habitat for the larvae— **Osmylus fulvicephalus** (Scop.)

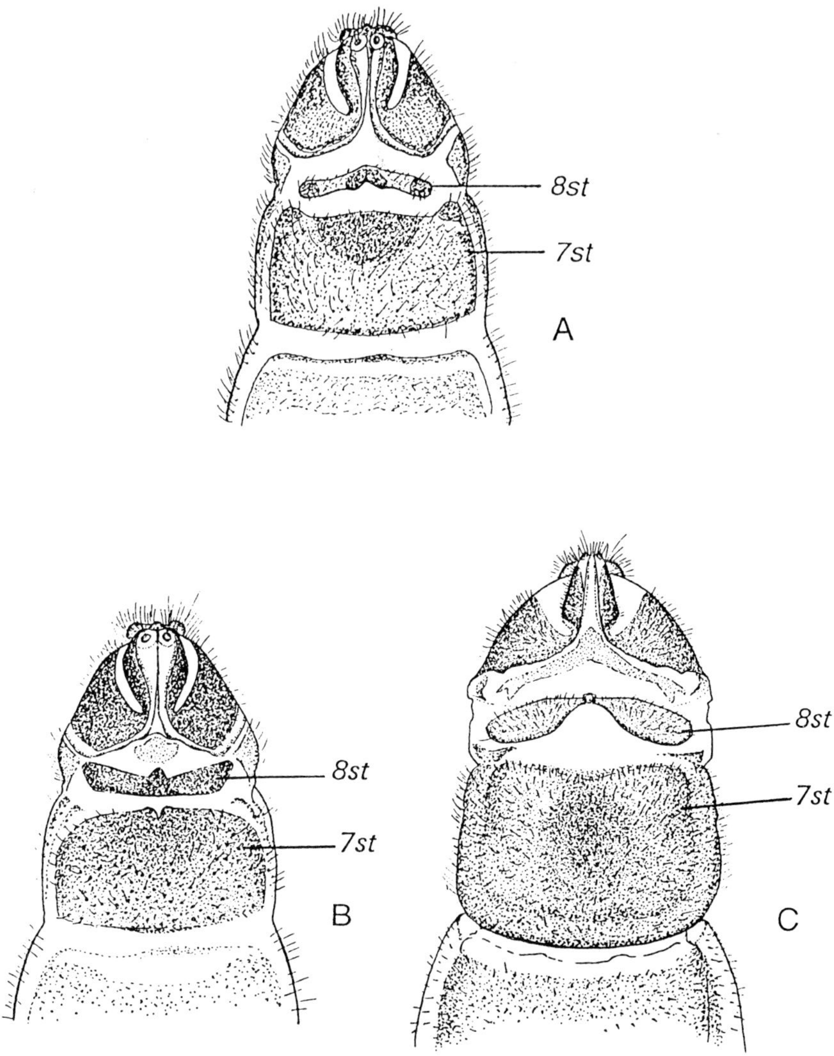

Fig. 5. External genitalia of female *Sialis*: A, *S. nigripes* in ventral view; B, *S. lutaria* in ventral view; C, *S. fuliginosa* in ventral view. 8*st*, 8th sternite; 7*st*, 7th sternite.

Family SISYRIDAE. Genus *Sisyra* Burmeister

Although the general characters are usually satisfactory for separating the three species of *Sisyra*, they are sometimes obscure in specimens preserved in fluid and are best seen in dry specimens mounted on pins. Therefore the external genitalia provide the most consistent characters for separating the adults. The sexes are easily recognised. Males have distinct claspers (Fig. 7A, B, C, D) which are absent in the females (Fig. 7E, F, G).

GENERAL CHARACTERS

1 Antennae are dark brown or black with pale yellow or yellowish-white tips. Wings are grey, almost colourless, with pale cross-veins and usually with only one cross-vein in the apical third of the wing, placed between R_1 and R_2 (Fig. 6A). Length of forewing 5–6 mm. Flight period May-August. Adults have been found chiefly near large rivers containing sponges on which the larvae live and feed—
Sisyra terminalis Curt.

— Antennae are dark throughout. Wings are a dingy-brown, blackish-brown or reddish-brown— **2**

2 The gradate cross-veins are not darker than the rest of the wings. Only one cross-vein is usually present in the apical third of the wing and this is usually placed between R_1 and R_3 (Fig. 6B), or there are no cross-veins in the apical region. Length of forewing 5–6 mm. Flight period May-September. Adults are usually found near the banks of lakes, ponds, canals and streams, all containing sponges—
Sisyra fuscata (Fabr.)

— The gradate cross-veins of the forewing and the membrane next to the cross-veins are distinctly darker than the rest of the wing, and the cross-veins appear to be thicker than the other wing-veins (Figs. 3B, 6C). One, two or three cross-veins are present in the apical third of the wing (Fig. 6C). Length of forewing 5–6 mm. Flight period June-August. Little is known about the ecology of this rare species but adults apparently occur not only near fast-flowing rivers, but also canals and lakes— **Sisyra dalii** McLach.

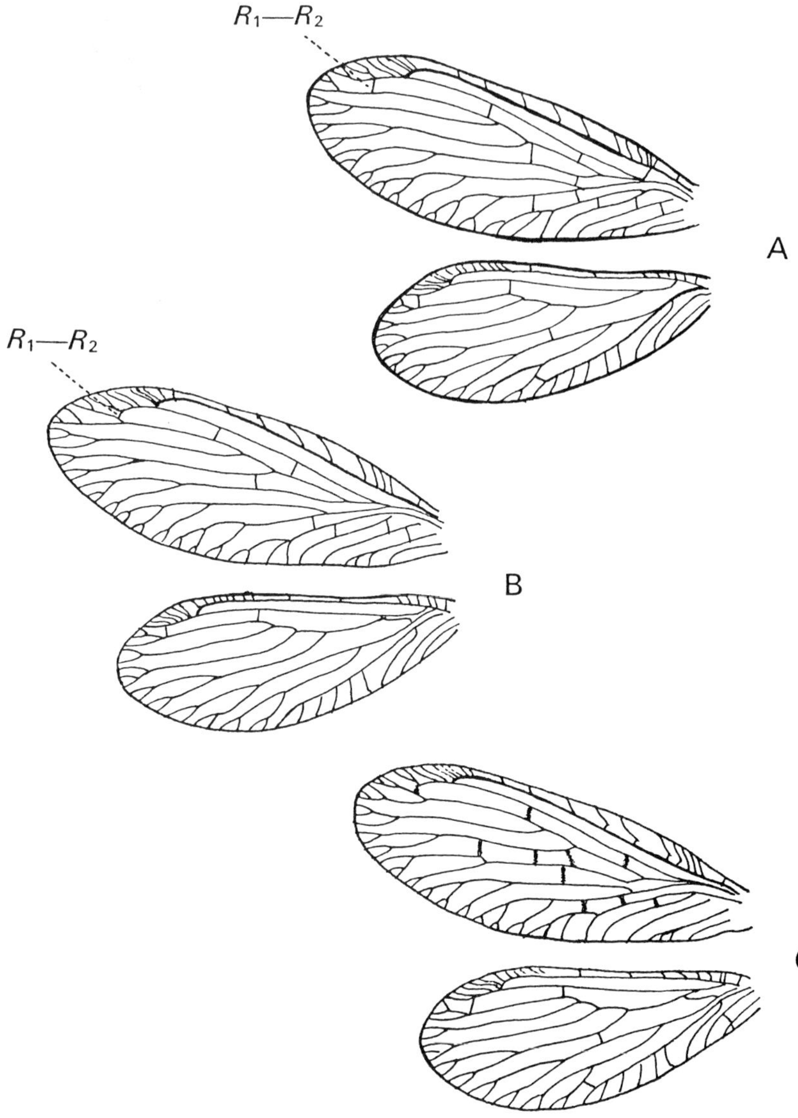

Fig. 6. Wings of *Sisyra* (length of forewing 5–6 mm): A, *S. terminalis*; B, *S. fuscata*; C, *S. dalii*.

MALE GENITALIA

1 Claspers are inwardly curved and taper to an acute apex (Fig. 7A, B)— **Sisyra fuscata** (Fabr.)

— Claspers are slightly convergent and truncate at the apex (Fig. 7C, D)— **2**

2 Claspers have sinuate margins; the lower margin ends in a small acute process which curves upwards and inwards, whilst the upper margin ends in three processes, a short stout spine, a long stout spine and a small tooth-like process (Fig. 7C)— **Sisyra terminalis** Curt.

— The upper margin of each clasper ends in a short inwardly-curved tooth below which are one or two blunt projections, with a slightly longer acute process near the lower margin (Fig. 7D)— **Sisyra dalii** McLach.

FEMALE GENITALIA

Although there are slight differences in the genitalia, especially in the form of the valve-like processes of the 9th sternite, these differences are not considered large enough to provide good characters for separating females of the three species (Fig. 7E, F, G). Figures are provided so that males and females can be recognised.

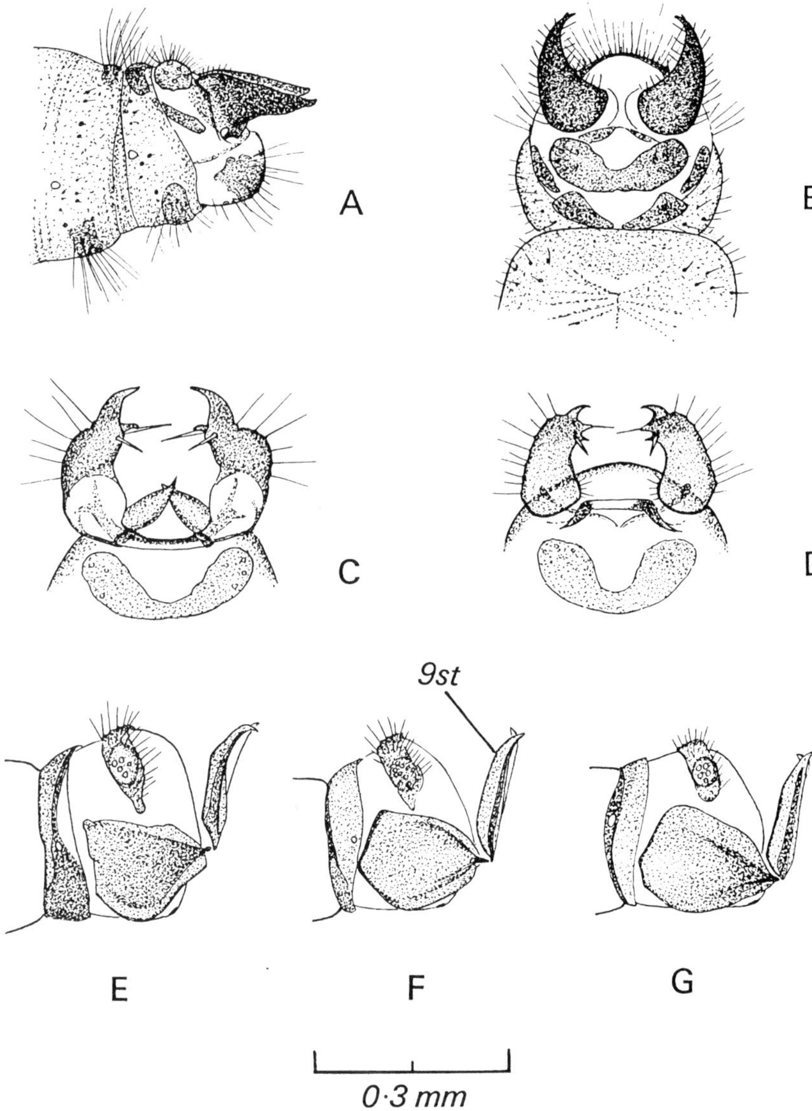

Fig. 7. External genitalia of *Sisyra*: A, male *S. fuscata* in lateral view; B, male *S. fuscata* in dorsal view; C, male *S. terminalis* in dorsal view; D, male *S. dalii* in dorsal view; E, female *S fuscata* in lateral view; F, female *S. terminalis* in lateral view; G, female *S. dalii* in lateral view. 9*st*, 9th sternite.

KEY TO LARVAE

The general characters given in the Introduction separate the Megaloptera and Neuroptera from all other orders of insects. The most useful characters are the fusiform shape of the body and the large "jaws" which are either large mandibles with sharp teeth for biting (Megaloptera), or modified mandibles and maxillae which fit together to form two forwardly-directed "spears" or "jaws" used for sucking the contents of the prey (Neuroptera). Some beetle larvae may be mistaken for larvae of Neuroptera, but they usually have a parallel-sided body and smaller jaws. Larvae of Gyrinidae (whirligig beetles) may sometimes be mistaken for *Sialis* larvae because they also have lateral abdominal gills. However, the gills are jointed in *Sialis* and the abdomen ends in a long tail, whilst the gills of Gyrinidae are unjointed and the abdomen ends in four hooks instead of a tail. Larvae of Trichoptera (caddis-larvae) may sometimes be mistaken for larvae of *Sialis* or *Sisyra* because some free-living and net-spinning species also have abdominal gills and prominent "jaws". However, the gills are never jointed in Trichoptera and the abdomen ends in a pair of legs, each carrying a curved claw.

As terrestrial larvae of Megaloptera and Neuroptera are often found near water and sometimes fall into the water, they may be confused with truly aquatic larvae of *Sialis* and *Sisyra*, and especially with the semi-aquatic larvae of *Osmylus fulvicephalus*. Therefore a key to families is given before the key to aquatic and semi-aquatic larvae.

KEY TO FAMILIES

1 Larvae have powerful mandibles armed with teeth and used for biting (Fig. 8A)— MEGALOPTERA **2**

— Larvae have apposed mandibles and maxillae which are hollowed out to form two tubes or suctorial jaws projected in front of the head (Fig. 8B)— NEUROPTERA **3**

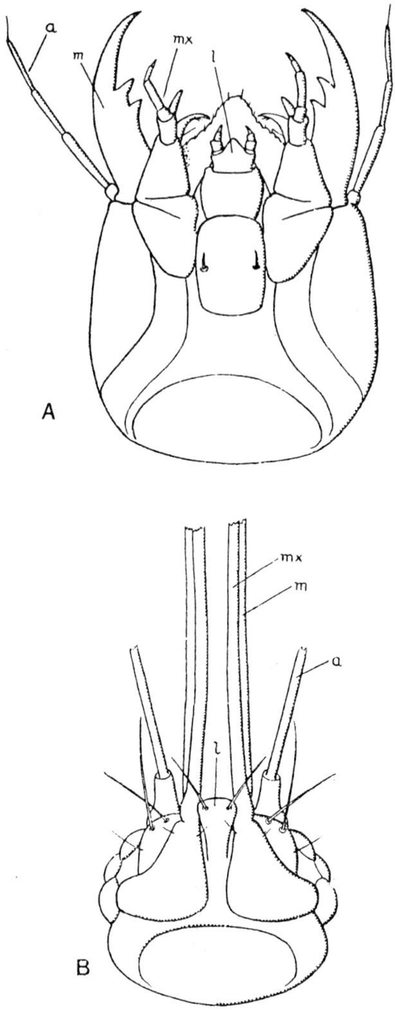

Fig. 8. Ventral view of larval heads of: A, *Sialis lutaria*; B, *Sisyra* sp. *a*, antenna, *m*, mandible, *mx*, maxilla, *l*, labium.
(Only the basal parts of the antennae, mandibles and maxillae are shown in *Sisyra*.)

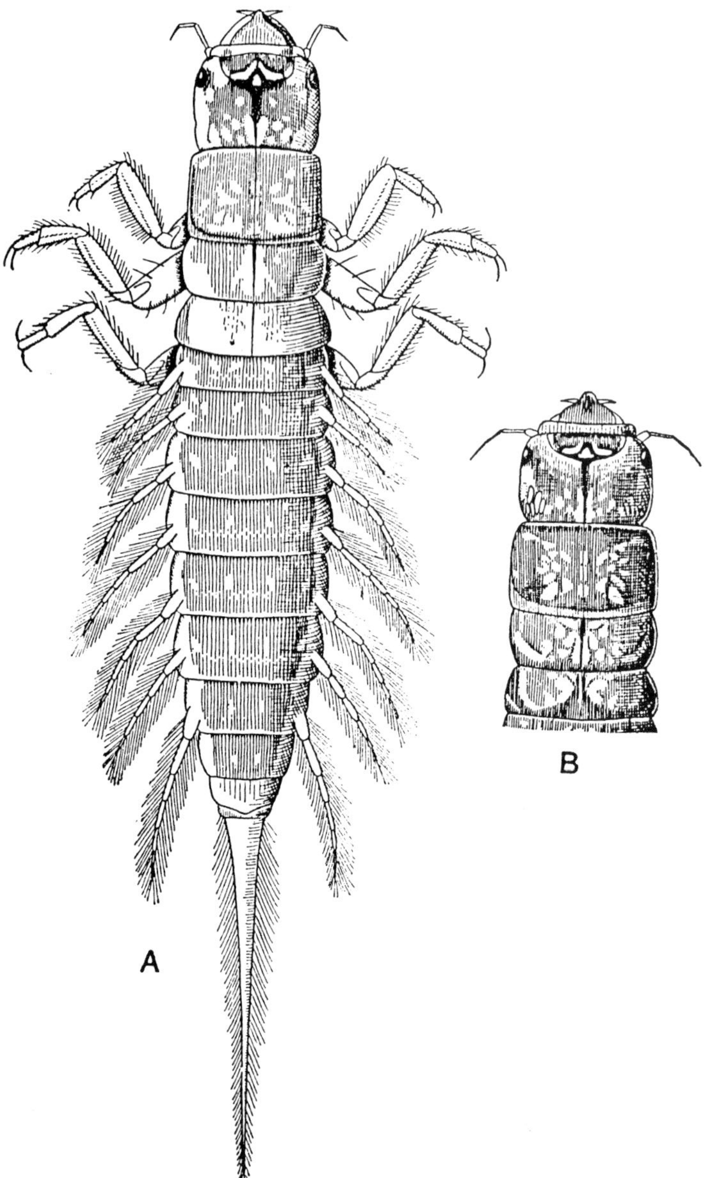

Fig. 9. A, larva of *Sialis fuliginosa*; B, head and thorax of larva of *S. lutaria* x 6

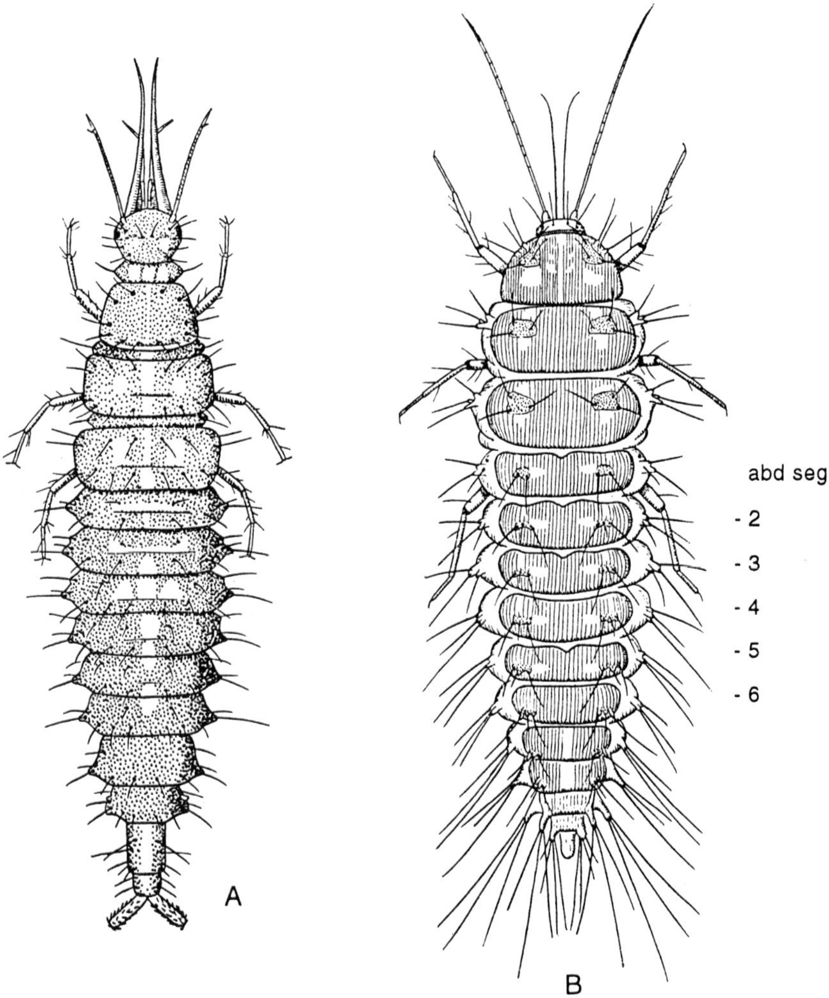

Fig. 10. Larvae of: A, *Osmylus fulvicephalus* (after Killington), x 7; B, *Sisyra fuscata* x 16. *abd seg*, abdominal segments (2–6).

2 Larvae are terrestrial and have no abdominal processes or appendages— RAPHIDIIDAE

— Larvae are aquatic and have a long tapering tail and seven pairs of segmentally-arranged jointed gills on the lateral margins of the abdomen (Fig. 9)— SIALIDAE, p. 28

3(1) Larvae are aquatic and live in or on sponges (Fig. 10B). There are seven pairs of segmentally-arranged jointed gills on the ventral surface of the abdomen (Fig. 11). The mouthparts are modified to form long slender sucking tubes (Fig. 8B) and each tarsus has only one claw with no empodium* next to the claw— SISYRIDAE, p. 31

— Larvae are terrestrial or live at the margins of streams, and have no gills. Each tarsus has two claws and an empodium* between the claws— **4**

* The empodium is a cushion-like structure which may be a tactile organ, or may be used to hold on to smooth surfaces.

4 Larvae are semi-aquatic, living in moss at the edge of streams. Mandibles and maxillae form long, slender, stylet-like tubes which project in front of the head and are slightly curved upwards (Fig. 10A). The body bears numerous short setae and the abdomen terminates in a pair of eversible finger-like processes which are covered with recurved hooks. Empodium is tapered with short fine hairs at the apex— OSMYLIDAE, p. 31

— Larvae are terrestrial. Jaws are prominent, caliper-like structures that curve inwards. The abdomen terminates in a pair of eversible pads, not covered with hooks— **5**

5 Jaws are long, slender and strongly curved. Empodium is trumpet-shaped. Body is usually clothed with long setae and there are usually prominent dorsolateral tubercles, bearing long setae— CHRYSOPIDAE

— Jaws are stout and only slightly curved. Empodium is short and pad-like. Body is rather sparingly clothed with fine and inconspicuous hairs and there are no prominent tubercles— HEMEROBIIDAE

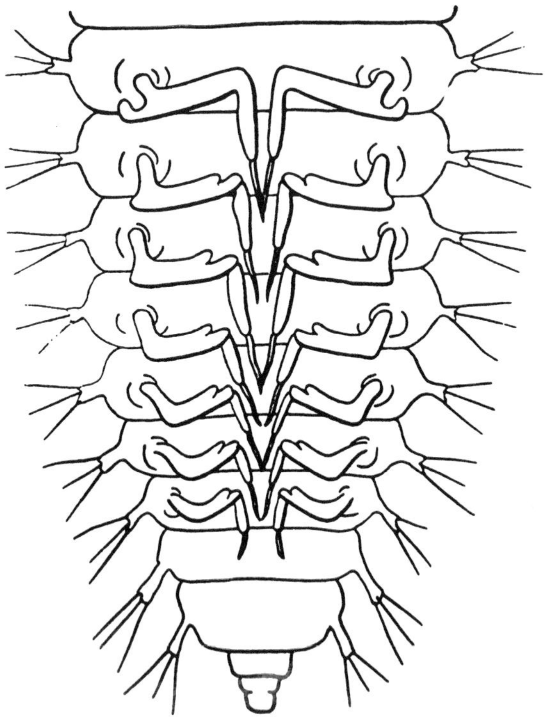

Fig. 11. Ventral view of abdominal segments and gills of larva of *Sisyra fuscata*, adapted from Lestage.

KEY TO SPECIES WITH AQUATIC OR SEMI-AQUATIC LARVAE

Family SIALIDAE. Genus *Sialis* Latreille

The earlier FBA publication by Kimmins (1962) used the following characters to separate two species: abdomen purplish-pink, head and thorax dark brown in *S. lutaria*; abdomen orange-yellow, head and thorax yellow-brown in *S. fuliginosa*. Unfortunately, the colour of the larvae is very variable and therefore these characters are unreliable. The characters used in the following key are the only morphological features that were found to be consistently different between species (Elliott, O'Connor & O'Connor 1979). They are applicable to instars 4 to 10, but may not always work for instars 2 and 3, and are definitely not applicable to the 1st instar which is unlike all the other instars.

The form of the mandibles, maxillae, labium, gular sclerite, abdominal gills and terminal process is very similar in the three species, but there are some differences in the number of spines and hairs on the legs. For example, the number of spines on the anterior face of the tibia of the foreleg is usually between five and seven in *S. lutaria*, between seven and nine in *S. fuliginosa*, and between twelve and fifteen in *S. nigripes*, but exactly the opposite is seen in some specimens.

1 Postoccipital sutures are distinctly arched and end at a short distance from the epicranial suture (Fig. 12A). A distinct dark band usually runs from the eyes in an oblique direction towards the posterior. Each abdominal tergum usually has a large median pale spot and two small lateral pale spots (Fig. 12D), but these spots may be fused in some specimens (Fig. 12E) or difficult to see in very dark specimens. As the life cycle usually takes two years, larvae are found throughout the year in ponds, lakes and sluggish streams and rivers where there is an abundance of silt. Length of larva, without terminal filament, is about 1 mm in the 1st instar and between 12 and 20 mm in the 10th and final larval instar— **Sialis lutaria** (Linn.)

— Postoccipital sutures are not arched and meet the epicranial suture directly or in the form of a groove (Fig. 12B, C). There is no dark band near the eyes. Each abdominal tergum has two pale lateral spots with no median spot (Figs 9A, 12F), but these spots may be difficult to see in pale specimens— **2**

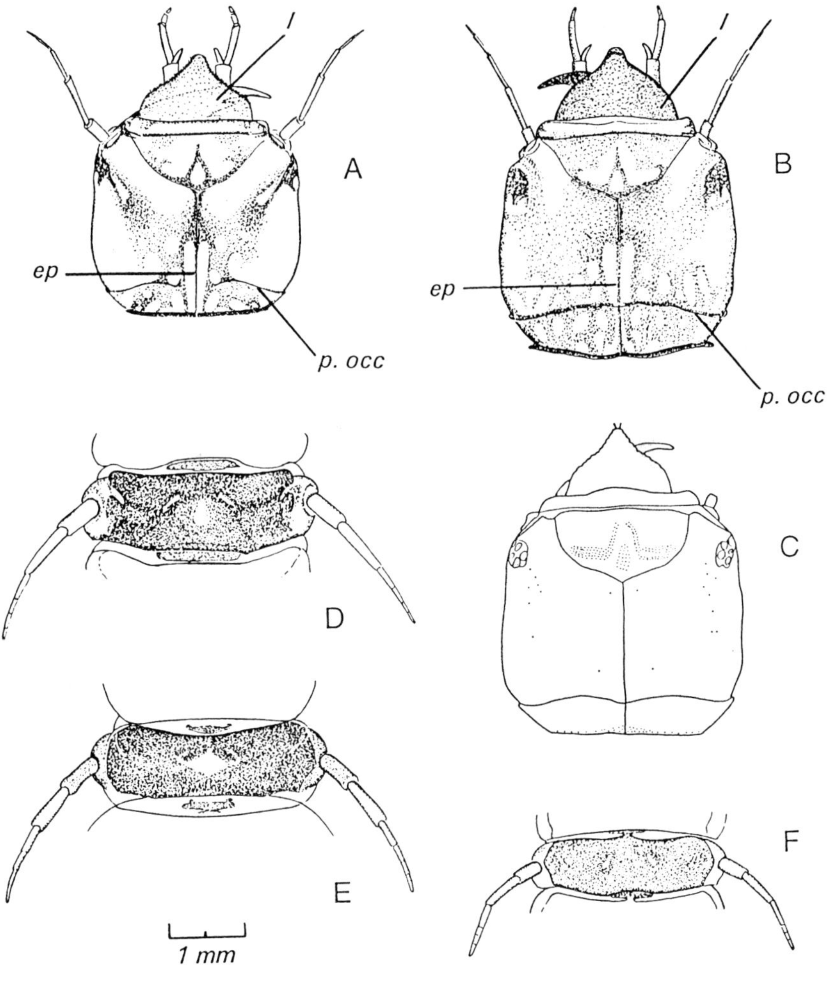

Fig. 12. Larvae of *Sialis*: A, head of *S. lutaria* in dorsal view; B, head of *S. fuliginosa* in dorsal view; C, head of *S. nigripes* in dorsal view; D, E, 2nd abdominal segment of *S. lutaria* in dorsal view, showing typical pattern and fused pattern; F, 2nd abdominal segment of *S. fuliginosa* in dorsal view. *l*, labrum; *ep*, epicranial suture; *p. occ*, postoccipital suture.

2 Head is dark brown with a distinct pattern throughout (Fig. 12B). Postoccipital sutures are not normally arched. Head and thorax are clearly darker than the abdomen in live specimens. The life cycle probably takes two years and larvae are found throughout the year in moderately fast streams and rivers. Length of larva, without terminal filament, is about 1 mm in the 1st instar and between 16 and 26 mm in the 10th instar— **Sialis fuliginosa** Pict.

— Head is yellow, almost without a pattern, apart from some markings on the frontoclypeus (Fig. 12C). Postoccipital sutures are slightly arched. Head and thorax are clearly lighter than the abdomen in live specimens. The life cycle is unknown but larvae are found chiefly in moderately fast streams and rivers (note that both adults and larvae have also been taken in Lough Ree, Ireland; see Elliott, O'Connor & O'Connor 1979). Length of larva, without terminal filament, is about 1 mm in the 1st instar and between 14 and 21 mm in the 10th instar— **Sialis nigripes** Pict.

Family OSMYLIDAE. Genus *Osmylus* Latreille

There is only one British species (Fig. 10A). The life cycle takes about one year and larvae can be found throughout the year in the moss bordering woodland streams. Larvae may be difficult to find from late September to March because they overwinter deep in the roots of the moss and are in a state of hibernation known as diapause. Length of larva, including jaws, is about 5 mm in the 1st instar, about 9 mm in the 2nd instar and about 15 mm in the 3rd and final larval instar— **Osmylus fulvicephalus** (Scop.)

Family SISYRIDAE. Genus *Sisyra* Burmeister

It has not been possible to construct a complete key to larvae because of the lack of specimens. The larva of *S. dalii* has not been found or described, except by Navás (1935) who gives a very poor figure of a larva which he believes to be that of *S. dalii*. Withycombe (1923) states that he had seen larvae of *S. terminalis* and *S. fuscata*, but he points out no differences of any kind between the two species. 1st- and 3rd-instar larvae of the relatively common *S. fuscata* have been found by several workers and described in detail by Withycombe (1923) and Killington (1936).

This situation has improved with the recent descriptions of the 2nd-instar larva of *S. fuscata* and the 1st-, 2nd- and 3rd-instar larvae of *S. terminalis* by Weissmair & Waringer (1994). These descriptions are used as the basis for the following key (see p. 32) to separate the two species.

FIRST-INSTAR LARVAE

(Unlike other instars, abdominal gills are absent; body length 0.5–1.0 mm)

1 Tarsal claw (for position, see Fig. 13A) has a semi-circular incision near its base (arrowed in Fig. 13B). On both the mesothorax and metathorax (for position, see Fig. 13A), there is only one pair of lateral protuberances, and each protuberance has only one seta (arrow in Fig. 13D)— **Sisyra fuscata** (Fabr.)

— Tarsal claw does not have a semi-circular incision (Fig. 13C). On both the mesothorax and metathorax, there are two pairs of lateral protuberances (arrows in Fig. 13E), and one pair has single setae whilst the other pair has two setae (Fig. 13E)— **Sisyra terminalis** Curt.

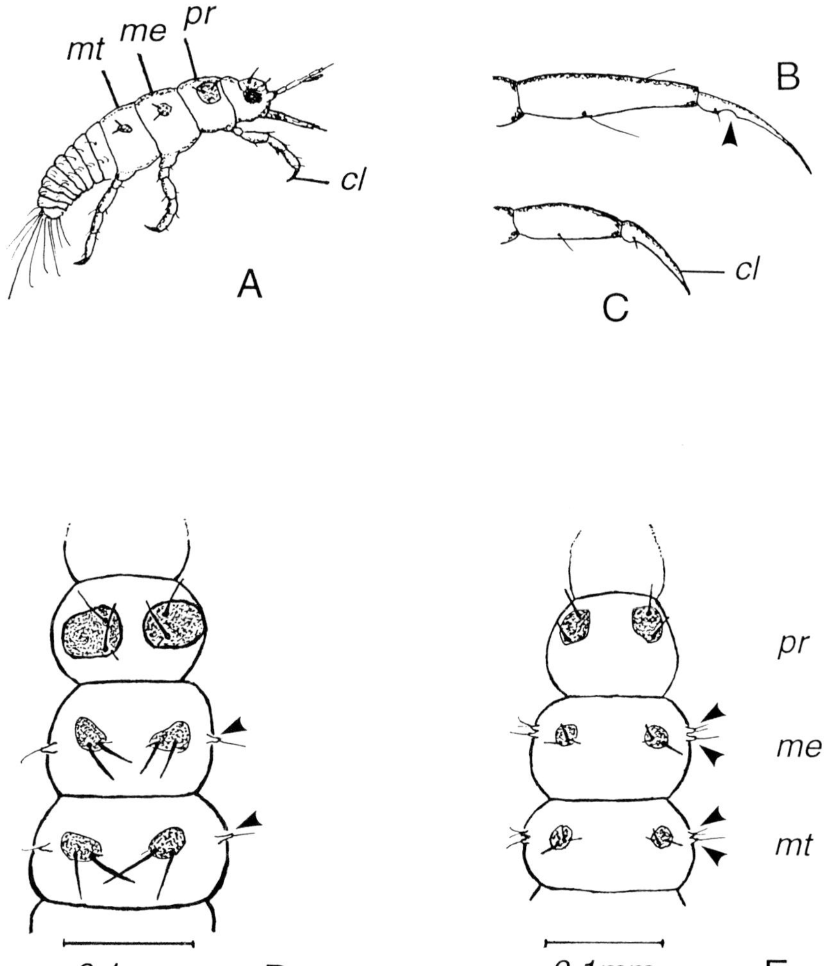

Fig. 13. 1st-instar larvae of *Sisyra*: A, lateral view of *Sisyra terminalis*; B, tarsus and claw of *S. fuscata*; C, tarsus and claw of *S. terminalis*; D, dorsal view of thorax of *S. fuscata*; E, dorsal view of thorax of *S. terminalis*. *pr*, prothorax; *me*, mesothorax; *mt*, metathorax; *cl*, claw.

SECOND-INSTAR LARVAE

(Abdominal gills are present; body length with mouthparts 1.2–2.4 mm)

1 1st pair of gills have only two segments with the basal segment curved, and the remaining gills have three segments with the basal segment bent towards the anterior and having a small apical projection (Fig. 14A)— **Sisyra fuscata** (Fabr.)

— All seven pairs of gills have three segments and all their basal segments are straight with no apical projections (Fig. 14B)— **Sisyra terminalis** Curt.

THIRD-INSTAR LARVAE

(Abdominal gills are present; body length with mouthparts 2.0–6.0 mm)

1 Setae triplets on dorsal surface of abdominal segments 2–6 (for position, see Fig. 10B) have evenly spaced setae, all pointing backwards (Fig. 14C). 1st pair of gills have only two segments with a projection on the basal segment (Fig. 14E). The latter projection also occurs on gills 2–5. Each antenna has 14–16 segments (Fig. 14G; this feature is the most difficult to use, especially when there are only 14 segments)— **Sisyra fuscata** (Fabr.)

— Setae triplets on dorsal surface of abdominal segments 2–6 have a median seta which points towards the anterior and is close to the lateral seta (Fig. 14D). 1st pair of gills have three segments and projections are absent from all gills (Fig. 14F). Each antenna has 11–13 segments (Fig. 14H)— **Sisyra terminalis** Curt.

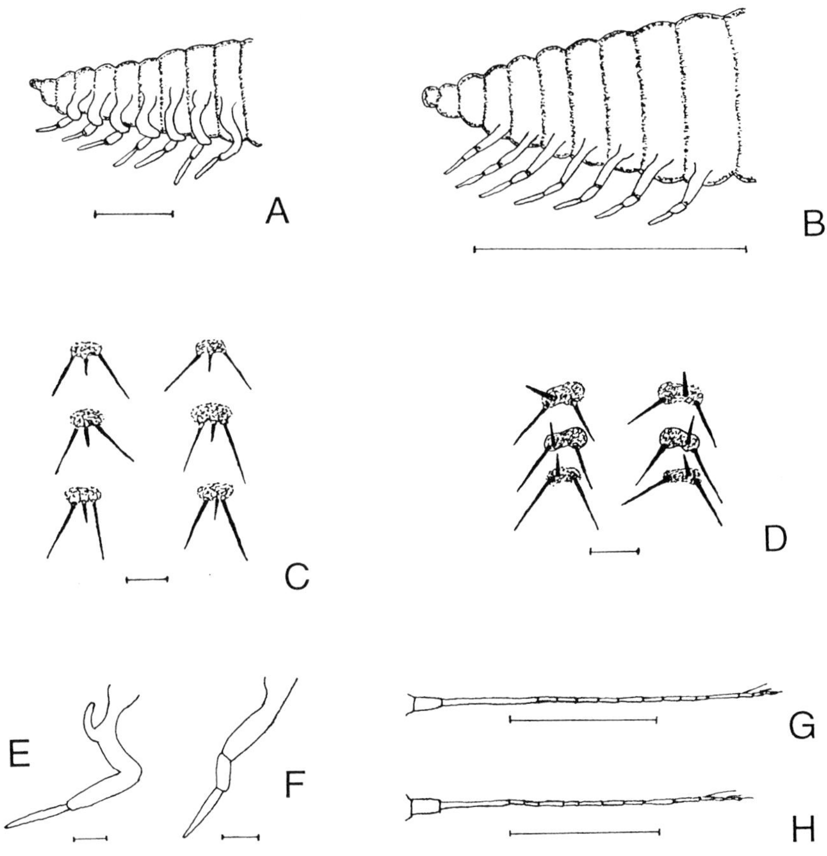

Fig. 14. 2nd-instar and 3rd-instar larvae of *Sisyra*: A, lateral view of abdomen of 2nd instar of *S. fuscata*; B, lateral view of abdomen of 2nd instar of *S. terminalis*; C, dorsal setal triplets of abdominal segments 2–4 of 3rd instar of *S. fuscata*; D, dorsal setal triplets of abdominal segments 2–4 of 3rd instar of *S. terminalis*; E, 1st abdominal gill of 3rd instar of *S. fuscata*; F, 1st abdominal gill of 3rd instar of *S. terminalis*; G, antenna of 3rd instar of *S. fuscata*; H, antennna of 3rd instar of *S. terminalis*. Scale lines: 0.5 mm for A, B, G, H; 0.2 mm for C, D, E, F.

LIFE CYCLES AND ECOLOGY

Genus *SIALIS*

The family Sialidae has a wide distribution, occurring in Europe, America, Asia Minor, Siberia and Japan. Six species, all in the genus *Sialis*, occur in Europe (Kaiser 1977; Aspöck, Aspöck & Hölzel 1978) but five additional species have now been recorded in far eastern Europe and the Caucasus (Vshivkova 1985). The last edition of this publication (Elliott 1977) included only two species, *Sialis lutaria* and *S. fuliginosa*, and whilst it was in press, two adults of a third species, *S. nigripes*, were discovered in Ireland. A re-examination of material in the British Museum (Natural History) showed that several specimens had been identified wrongly and were actually *S. nigripes* (Barnard 1977). The earliest records were from 1867 in England and 1914 in Ireland, so this species has been an undetected member of the British and Irish fauna for some time. Earlier records are given in Barnard (1977), but there are several additional records, including central Scotland (O'Connor & Norton 1977, 1978; Speight 1977; Elliott, O'Connor & O'Connor 1979; Mackey & Ham 1980; Fozzard & Clelland 1981; O'Connor & O'Grady 1990). Many workers have studied the general biology of *S. lutaria* but comparatively little is known about *S. fuliginosa* and *S. nigripes*. All three species sometimes occur at the same site and there is some evidence to suggest that adults of *S. lutaria* emerge first, followed by *S. fuliginosa* and *S. nigripes* (Kaiser 1950; Mackey & Ham 1980; Stroot 1986).

Sialis lutaria

The large number of studies on the internal anatomy and physiology include work on the development of the embryo (Du Bois 1936), the circulatory system of adults and larvae (Selman 1965), the morphology and behaviour of the blood cells (Selman 1960a,b, 1962), the blood composition, ionic regulation and water balance in the larva (Beadle & Shaw 1950; Shaw 1955a,b; Sutcliffe 1962, 1963), the excretion of ammonia by the larva (Staddon 1955), and the physiological changes during pupation (Geigy 1948; Geigy & Ochsé 1940a,b; Ochsé 1944; Geigy & Rahm 1951; Rahm 1952; Selman 1960a).

Sialis lutaria is a fairly widespread species throughout the whole of Europe, except the southeast (Vaňhara 1970). The larvae live in ponds, lakes and sluggish parts of streams and rivers where there is an abundance of silt. They are often numerous in the benthos, especially in humic, acid lakes and ponds (Berg & Petersen 1956; Matthey 1971; McLachlan & McLachlan

1975). The name *lutaria* is derived from the habitat of the larva (Latin *lutum* = mud). Detailed studies of the life cycle have been made in Denmark (Berg 1938; Kaiser 1950, 1961, 1977; Iversen & Thorup 1987; Dall 1987, 1989), France (Giani & Laville 1973), Germany (Seitz 1940) and Switzerland (Du Bois & Geigy 1935; Geigy & Du Bois 1935; Geigy 1937; Geigy & Grobe 1958). Early accounts of the life cycle in Britain (McLachlan 1868; Killington 1929; Syms 1935) are brief, but a general account for this species in Windermere (English Lake District) is provided here and a detailed study has been made on a population in a drainage channel in southeast England (Brooker 1979). The life cycle usually takes about two years from egg to adult but a longer period of three years has been recorded in lakes at high altitudes in Switzerland (Geigy & Grobe 1958) and the French Pyrenees (Giani & Laville 1973), and a shorter period of one year in Denmark (Iversen & Thorup 1987; Dall 1989).

The flight period in Britain is from late April to the end of June but adults have occasionally been taken in July in the south of England (Killington 1926). Their flight is very slow and laboured, and when disturbed, they usually run rather than fly. The flight period starts later with increasing latitude, probably due to temperature differences. It is from April to May in Switzerland (Du Bois & Geigy 1935), April to June in Belgium (Stroot 1986), April to July in Czechoslovakia (Vaňhara 1970), May to June in Denmark (Kaiser 1950) and southern Finland (Meinander 1962). In Norway, adults of this species are most frequently found in June and July (Solem 1973) and were caught in light traps from May to mid-June (Anderson & Greve 1975). It is notable that in the latter study, 97% of the adults taken in the light trap were males, suggesting that they were more readily attracted to light and/or had a wider flight range than females.

Females are much larger than males; e.g. the means and ranges for body length in a hundred adults of each sex from various localities in the English Lake District were 10 mm (7–13 mm) for males and 13 mm (11–16 mm) for females. Although the adults have biting mouthparts, there are no definite records of them feeding. Du Bois & Geigy (1935) found pollen on the head and thorax, and therefore concluded that the adults visit flowers to feed. Kaiser (1961) observed females apparently looking for food on flowers of cow parsley *Anthriscus silvestris* (L.) Hoffm. Similar behaviour was observed in two North American species, *S. rotunda* Banks and *S. californica* Banks, by Azam & Anderson (1969) but these authors found that the alimentary canal was not developed and therefore feeding was doubtful. Adults of another North American species, *S. cornuta* Ross, fed on sugar solution in the laboratory and were kept alive for up to one week (Pritchard & Leischner 1973). The life of the adults is short, usually one week and rarely longer than two weeks. Dead or dying adults frequently fall into the water and are eaten by fish, some adults are eaten by birds and spiders, and dead males are sometimes found in a mummified condition clinging to plants (Du Bois & Geigy 1935).

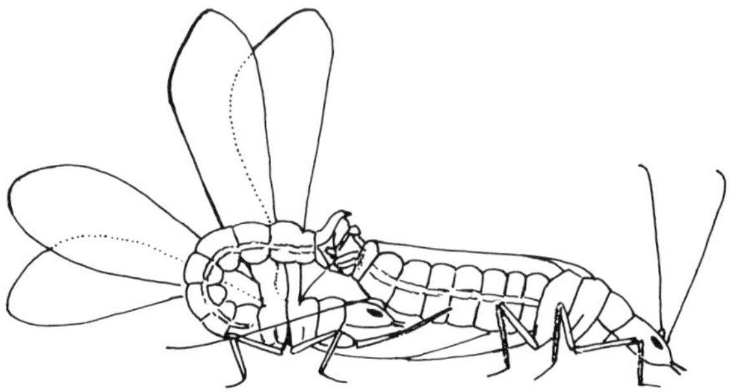

Fig. 15. Mating in *Sialis lutaria*; male is on the left, female on the right (cf. Plate 1).

Adults are most active on warm, sunny days, especially in the morning. Large females are sometimes seen in flight pursued by one or more smaller males. Mating always occurs on the ground, usually in the morning, whereas oviposition can occur throughout the day from mid-morning (about 11.00 hours) to evening. The female attracts the male by producing a scent for which the receptor is on the male antennae (Geigy & Du Bois 1935), and often a female is surround by several males. Courtship usually starts with one male confronting the female and both vibrating their antennae for a few seconds. The female then walks a short distance with the male following. When the female stops or slows down, the male vibrates his antennae and mouthparts over the hind legs, wing tips and end of the abdomen of the female. The antennae and notched labrum of the male are thought to be receptor organs (Geigy & Du Bois 1935). Rupprecht (1975, 1983) found that males and unmated females are able to find each other and communicate by generating vibration-signals. These are produced by rhythmic movements of the abdomen in both sexes and are transmitted through the legs and the ground to receptors on the legs of the opposite sex. If the female is ready for copulation, the male forces up the end of her abdomen with his head and moves underneath. He then turns his abdomen under one pair of wings and curls it over dorsally so that the genitalia meet (Fig. 15 and Plate 1). The pair remain in copulation for several minutes and a large white spermatophore is

Plate 1. Eggs and mating in *Sialis lutaria*. (Photo: J. M. Elliott).

transferred from male to female. After one or two twists of the abdomen, the male disengages, leaving the spermatophore held ventrally by the genitalia of the female. After walking a short distance, the female spreads her legs wide apart, curls her abdomen under her thorax and chews on the spermatophore. The female copulates only once.

Fig. 16. Eggs of *Sialis lutaria*; length with micropyle about 0.8 mm.

Soon after feeding on the spermatophore, the female searches for an oviposition site. The eggs are usually laid on the stems and leaves of plants overhanging the water but are occasionally found on stones, tree trunks and bridges. Each egg is brown, cylindrical, with a short white, cylindrical micropylar knob on the upper end (Fig. 16). The eggs are arranged in a single layer and at a slight angle to the surface on which they are attached (Plate 2). The female lays the eggs in rows and one female can lay between 300 and 900 eggs in a single mass (Du Bois & Geigy 1935; Seitz 1940; Kaiser 1961). Sometimes a second, smaller mass of eggs is laid after a rest-period of about 6 days. Lestage (1920a) gives a detailed description of oviposition behaviour in *S. lutaria*.

The eggs are frequently parasitised by a minute hymenopteron which was named *Oophthora semblidis* by its discoverer Aurivillius (1898) and later called *Pentarthron semblidis* by Lestage (1919) and *Trichogramma evanescens* Westw. by Kryger (1919), Lestage (1920a) and Du Bois & Geigy (1935). From his detailed study of the parasite, Salt (1937) concluded that the correct name was *Trichogramma semblidis* (Aurivillius). Salt collected nearly a quarter of a million eggs of *S. lutaria* and found that about 0.6% of them were attacked by the parasite. Much higher percentages have been reported for this species parasitising eggs of North American species, with values of 14% for *S. rotunda*, 37–65% for *S. californica* (Azam & Anderson 1969), and 78% for *S. cornuta* (Pritchard & Leischner 1973). It is easy to recognise eggs that were parasitised because the emerging parasite bites a small round hole whereas the *Sialis* larva makes a longitudinal slit on hatching.

The development of the embryo was described by Du Bois (1936) and Seitz (1940). The eggs usually hatch at night and nearly all larvae from one

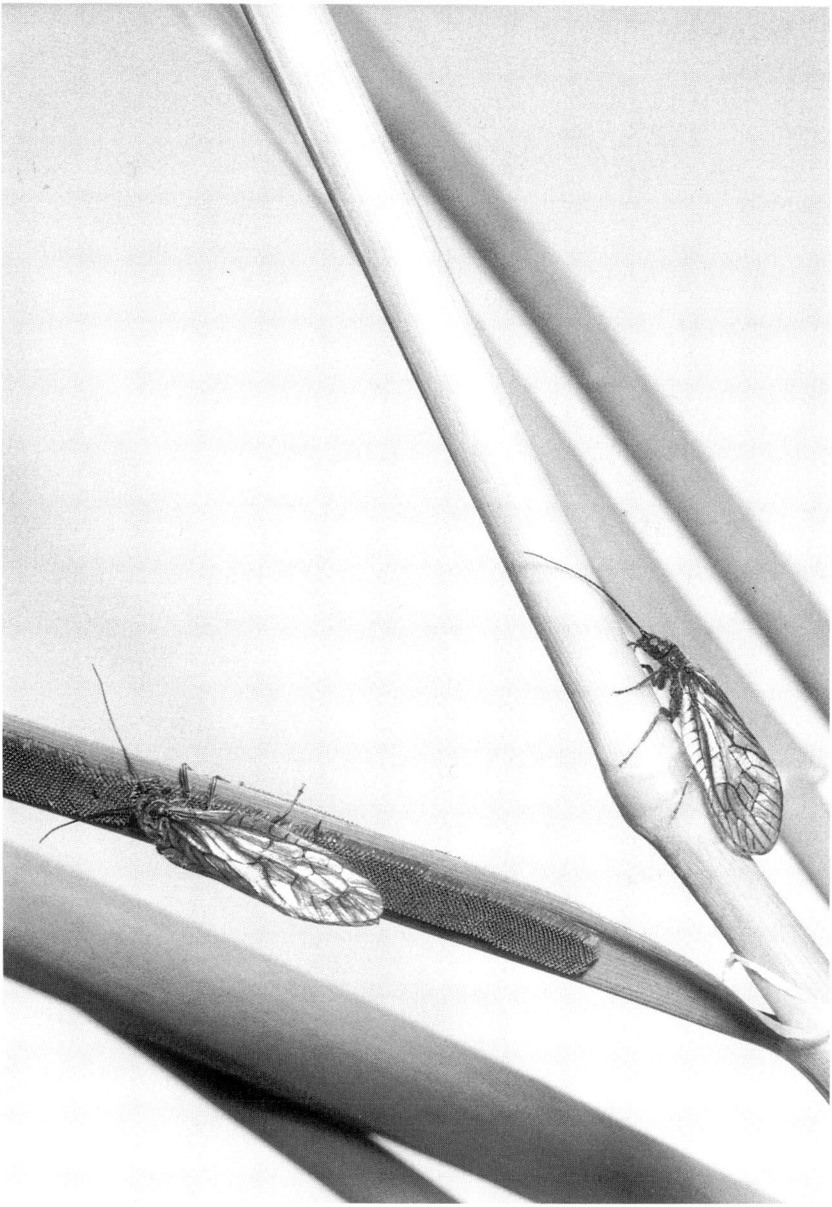

Plate 2. Eggs of *Sialis lutaria*, on reed-grass. (Photo: J. M. Elliott).

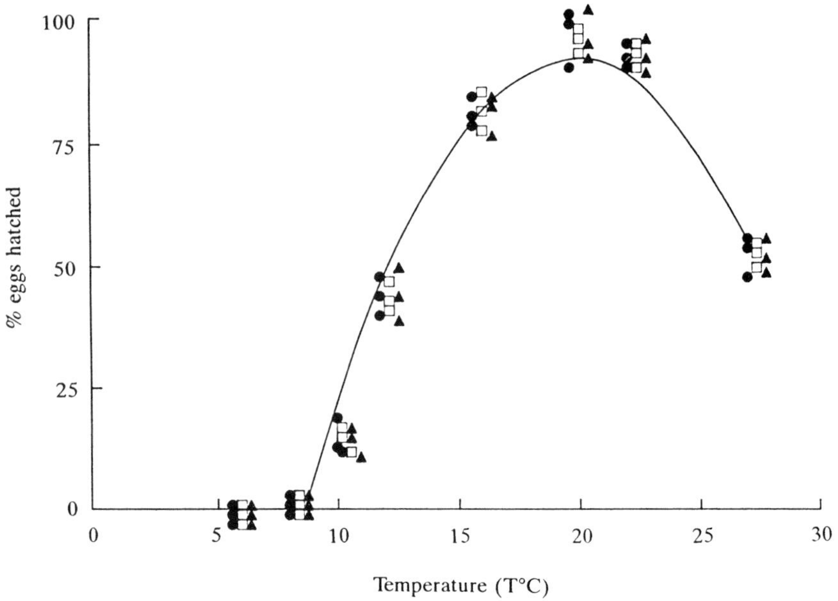

Fig. 17. Relationship between temperature (T °C) and percentage of eggs hatching for *Sialis lutaria* from Windermere (●), Black Beck (□) and the Lancaster Canal (▲).

egg mass emerge simultaneously. Most eggs hatch in May and June but a few may still be hatching in early July. An egg burster on the cuticle of the embryo is used to cut a slit through the egg shell. The 1st-instar larvae fall into the water or onto the ground and then crawl into the water. Development time varied from about 7 days at 24°C to about 20 days at 15°C and few eggs developed at temperatures below about 10°C (Seitz 1940). Kaiser (1961) obtained a mean incubation time of 17.5 days in the range 13.0–15.5°C.

Elliott (1995) provides the most detailed information on egg hatching over a wide range of temperatures (5.9–27.2°C). The eggs incubated at eight temperatures came from three populations; from a lake (Windermere), from a stream (Black Beck, the main inflow of Esthwaite Water in northwest England), and from a canal (Lancaster Canal in northwest England). The percentage of eggs hatching varied considerably with temperature but there were no significant differences between eggs from the three populations (Fig. 17). No eggs hatched at 5.9°C, very few eggs hatched at 8.2°C, few eggs (<18%) hatched at 10.0°C, but hatching was high at the remaining temperatures. The relationship between hatching success and temperature was well described by a second-degree polynomial and 98% of the variation in hatching success could be attributed to temperature variations. The

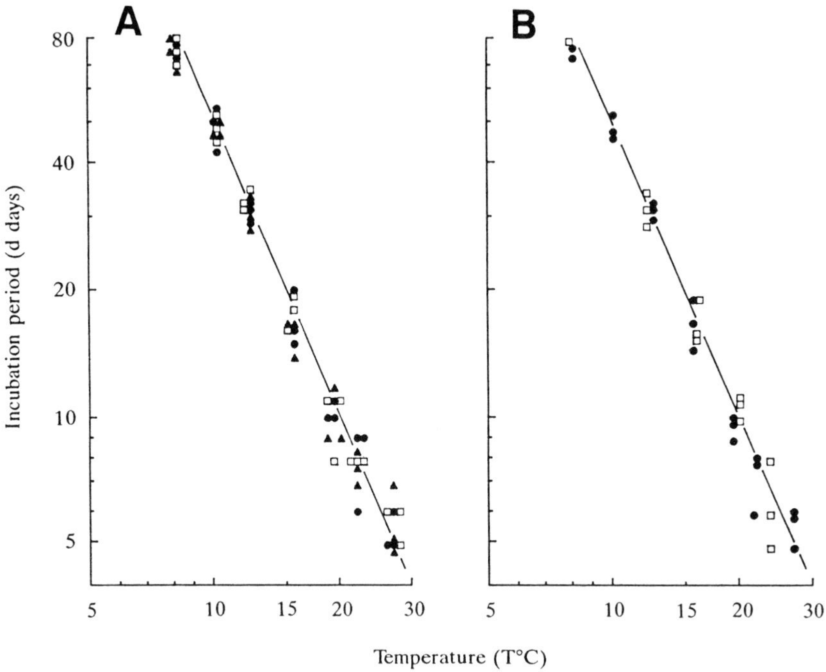

Fig. 18. Relationship between temperature (T °C) and incubation period (d days) for: A, *Sialis lutaria*; B, *S. fuliginosa*. Linear regression lines are fitted but data are presented on log-log scales (symbols are the same as those used in Figs 17 and 22).

curvilinear model indicated that the optimum temperature for hatching was 20.0 °C and that at least 50% of the eggs hatched in the range 12.2–27.5 °C. It will be shown later that these values differ from those for *S. fuliginosa*. The relationship between the incubation period (d days between oviposition and the date on which 50% of the eggs hatched) and temperature (T °C) was linear on logarithmic scales and was well described by a negative power function (Fig. 18A). There were no significant differences between eggs from the three populations (note also that the relationship was not significantly different from that for *S. fuliginosa*). Variations in temperature could account for over 98% of the variation in the length of the incubation period.

1st-instar larvae are between 0.7 mm and 2.0 mm long and are active swimmers, with long hairs on the legs, abdomen and gills (Fig. 19). Their

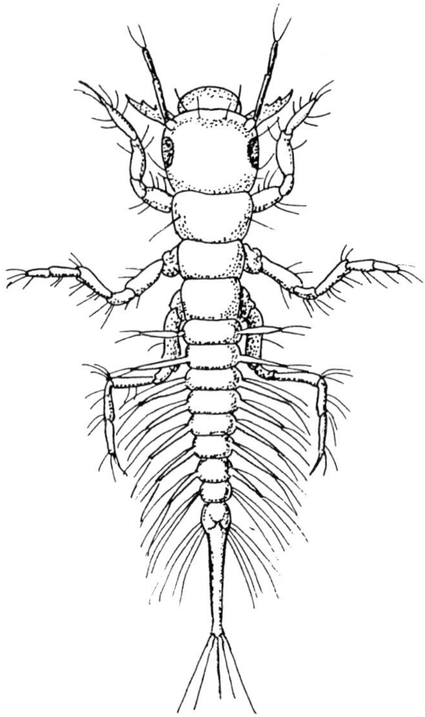

Fig. 19. 1st-instar larva of *Sialis lutaria*; overall length 0.7–2.0 mm.

appearance is very different from the other larval instars (see Fig. 9, p. 24), and both Lestage (1919) and Bertrand (1949) have compared the morphology of the two larval forms. As the 1st instars are essentially planktonic, they disperse rapidly from the comparatively limited area where newly-hatched larvae enter the water. They are positively phototactic whereas all other larval instars are negatively phototactic. The 1st instar moults after 3–5 days and the negatively phototactic 2nd instar descends to the bottom and burrows in the mud.

There are ten larval instars, and this appears to be the usual number for most species of *Sialis*, but a North American species, *S. cornuta*, pupates at any instar between seven and ten (Pritchard & Leischner 1973). Larvae of later instars of *S. lutaria* can be induced to pupate by experimental and surgical techniques, but the percentage success increases with the instar number (Selman 1960a). The size of each larval instar varies slightly from one locality to another and there is a marked sexual dimorphism in size in the

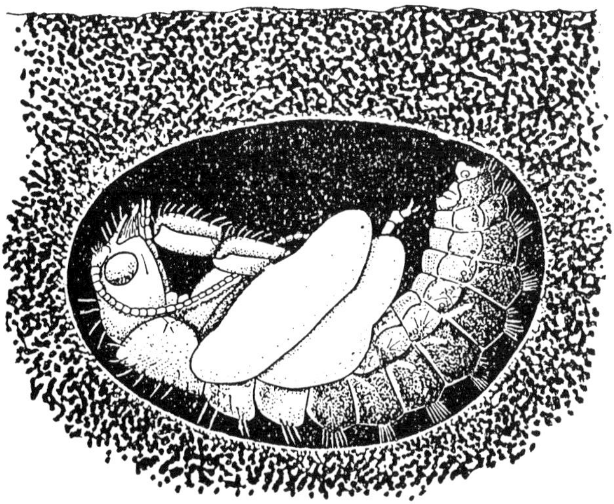

Fig. 20. Pupal chamber and pupa of *Sialis lutaria*.

later instars. For example, ranges for body length including the terminal filament are usually 14–22 mm in male larvae and 18–26 mm in female larvae. Although body length can be used to separate the ten instars, a more precise character is the width of the head capsule (Giani & Laville 1973).

Most larvae are in the 8th or 9th instar by the end of their first year (May or June) and in the 10th instar in autumn (September or October). They usually pass their second winter in the 10th instar and finally leave the water for pupation (Fig. 20) in late March, April, May or the beginning of June. Such a life cycle is seen in Windermere (Fig. 21A), and in an English drainage channel (Brooker 1979). Therefore the larval stage usually lasts about two years but a longer period of three years has been found in lakes at high altitudes, with the larvae in only the 5th or 6th instar at the end of their first year (July or August) and in the 8th or 9th instar at the end of their second year (Geigy & Grobe 1958; Giani & Laville 1973). In marked contrast, the life cycle lasted only one year with very rapid growth in autumn in the Danish River Suså (Iversen & Thorup 1987), and with rapid growth in July in the Danish Lake Esrom (Dall 1989).

The larvae are carnivores and their predominant food organisms are chironomid larvae and oligochaetes in the larger larvae (Griffiths 1973; Giani & Laville 1973), benthic crustaceans in the smaller larvae (Giani & Laville 1973) and micro-organisms and detritus in the 1st-instar larvae (Du Bois & Geigy 1935). Older larvae (instars 7 to 10) sometimes eat younger larvae. Food organisms are usually swallowed whole but larger prey are manipulated

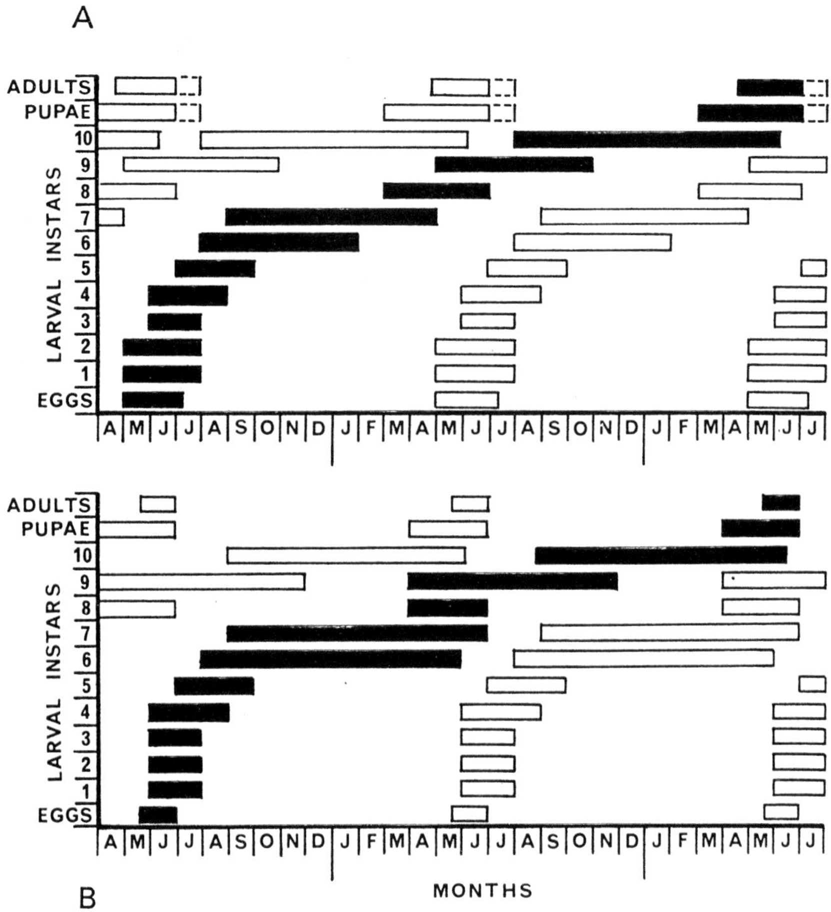

Fig. 21. Life cycle of: A, *Sialis lutaria* in Windermere; B, *Sialis fuliginosa* in the Walla Brook on Dartmoor and Wilfin Beck in the English Lake District. Each horizontal bar indicates the months in which the various stages of the life cycle occur.

into the foregut with the forelegs and mandibles, and then bitten into two pieces. The prey is retained in the foregut by proventricular teeth whilst enzymes from the midgut initiate breakdown of the food into fine particles which pass into the midgut. Digestion chiefly occurs in the foregut and undigested material passes rapidly through the midgut to form faecal pellets. Pritchard & Leischner (1973) found that the mean minimum time between ingestion of prey and evacuation of the first faecal pellet in *S. cornuta* was

106 h at 5 °C, 29 h at 12.5 °C and 22 h at 17.5 °C. *Sialis* larvae are eaten by other carnivorous invertebrates such as larvae of Odonata (Griffiths 1973), and also by fish such as brown trout *Salmo trutta* L. (e.g. Macan 1966; Giani & Laville 1973).

Larvae are sometimes found with internal parasites. A new species of hymenostome ciliate, *Tetrahymena sialidos* sp. nov., was found in the haemocoele of larvae from a lake population in mid-Wales, 40–70% of the larvae being infected (Batson 1985). Larvae did not die until 11–12 months after initial infection and the mean density of ciliates attained a density of 55,000 per host. The death of the host larvae and release of the ciliates coincided with the appearance of a new generation of host larvae in which the prevalence of infection increased rapidly. Juveniles of the acanthocephalan *Neoechinorhynchus rutili* (Müller) have also been found in the haemocoele of the larvae, the vertebrate host being brown trout *Salmo trutta* L., which fed on the *Sialis lutaria* larvae (Lassiere 1988). Larvae of a cestode species, *Proteocephalus* (probably *P. filicollis* (Rudolphi)), were found in up to 40% of larvae in a river in southwest England (Kennedy, Nie & Rostron 1992). The most frequent hosts for adults of this cestode were probably sticklebacks *Gasterosteus aculeatus* L. Very little is known about the long-term effects of these parasites on their hosts, but Batson (1985) discusses their regulatory influence on host population dynamics.

The larvae occur in the littoral, sub-littoral and sometimes the profundal zones of lakes and have been recorded at depths down to 7 m in Lake Gribsø (Berg & Petersen 1956), 12 m in Windermere (Humphries 1936; Macan & Worthington 1972) and Lake Esrom (Berg 1938), about 13 m in the Plöner lakes (Lundbeck 1926), 15 m in lakes Fure and Hald (Wesenberg-Lund 1917), Lake Lille-Jonsvann (Solem 1973) and Lake Port-Bielh (Giani & Laville 1973), and 20 m in Sempachersee (Du Bois & Geigy 1935). All instars do not have the same distribution pattern on the bottom. As the larvae grow, their maximum numbers are found at progressively greater depths. The smaller larvae in their first year are therefore most abundant in the littoral, whereas the larger larvae in their second year are most abundant in the sub-littoral or profundal zones, except when they migrate inshore before pupation in spring (Du Bois & Geigy 1935; Berg & Petersen 1956; Giani & Laville 1973). Migration traps at a water depth of 0.5 m in Lake Esrom caught 126 final-instar larvae in traps with the mouth facing away from the shore and only 7 larvae in traps in the reverse position; immigration to the shore occurred in April and May (Dall 1987).

10th-instar larvae leave the water in late March, April, May or early June, the exact time varying from one locality to another. Larvae usually move onto land in the evening and early hours of the night, and often move some distance from the water's edge before they find a suitable site for pupation.

Miall (1895) describes how a larva climbed a concrete wall, passed through a thicket of *Cotoneaster* and finally pupated in a flower bed that was six yards from the water. The tracheal gills of the larva adhere to the body instead of projecting outwards and their contents are retracted to form a small lump at the base of each gill.

Pupation usually occurs in fairly damp loose soil or in vegetable debris in the middle of clumps of sedge (Killington 1926). Each larva digs a pupal chamber about 1 cm below the soil surface and within a day it has moulted into a pupa. No cocoon is spun in the pupal chamber and the pupa has the legs and rudimentary wings free. The body is curved inwards and the abdomen has spines which keep the pupa free from direct contact with the walls of the chamber. In Fig. 20, the pupa is shown lying horizontally in a closed chamber just below the soil surface, but Jacques (1974) notes that the pupae lie vertically, head uppermost, in open chambers. In his detailed study of pupation, Seitz (1940) shows that there are actually three types of pupal chamber: an open-mouthed chamber in which the larva lies vertically, a J-shaped chamber which is open at one end with the larva lying vertically in the closed limb, and a closed chamber as in Fig. 20. He found between 30 and 40% of the pupae in closed chambers which are made only in firm soil. The pupal stage requires a total of 72.5 degree-days above a base temperature of 6.6°C (Kaiser 1961); e.g. it lasts 30 days at 9.0°C, 20 days at 10.2°C, 10 days at 13.9°C and 5 days at 21.1°C. Emergence of the adult usually occurs soon after dawn in late April, May, June and occasionally July. If the pupal chamber is open, the final moult occurs in the chamber, where the cast pupal skin is left, but if the chamber is closed, the pupa digs its way out and the final moult occurs on the soil surface.

The life cycle is thus complete and usually takes two years. A summary of the two-year cycle in Windermere is shown in Fig. 21A. As the actual months in which each instar occurs will vary from one locality to another, this figure can only serve as a guide. Annual production estimates are available for a few populations, with values of 0.51 g dry weight m^{-2} in a Pyrenean lake (Giani & Laville 1973), 0.98–1.17 g dry weight m^{-2} in a drainage channel in southern England (Brooker 1979), 0.16-2.49 g ash-free dry weight m^{-2} in the Danish River Suså (Iversen & Thorup 1987) and 0.1-1.2 g dry weight m^{-2} in the Danish Lake Esrom (Dall 1989). Annual production to mean biomass ratios were 1.6–1.8 in the drainage channel, 3.0–7.3 in the River Suså and 4.6–7.1 in Lake Esrom.

The common name of the adult *Sialis* is the alder or orl fly, and an imitation of the adult is used by fly-fishermen during the flight period from late April to June. Adults are often found in the stomachs of trout and are usually moribund adults that have fallen into the water after mating or oviposition. As they are poor fliers, some adults may become fatigued at some distance from the land, especially in a strong wind, and may then fall or be blown onto the water surface.

Sialis fuliginosa

Nothing is known about the internal anatomy and physiology of *S. fuliginosa* and there is little information on its ecology. This species occurs throughout Europe (Vaňhara 1970). The name *fuliginosa* (from Latin *fuligo*, soot or black paint) probably refers to the wings, which are blacker than those of *S. lutaria*. The larvae are limited to moderately fast streams and the upper reaches of rivers. As larvae of *S. lutaria* are sometimes found in rivers, there may be some places where the two species occur together, but nothing is known about any possible competition between them.

There are some detailed observations on the life cycle in Denmark (Kaiser 1950, 1961), and the following account is based on this work and some additional observations made in two small stony streams, the Walla Brook on Dartmoor in southwest England and the Wilfin Beck in the English Lake District. These streams are described in detail by Elliott (1967a, 1973). The life cycle takes two years (Fig. 21B).

In Britain, the flight period starts slightly later than that of *S. lutaria* and usually lasts about six weeks in May, June and occasionally July. Flight periods recorded elsewhere in Europe are mid-May to early July in Belgium (Stroot 1986), mid-May to end of June in Denmark (Kaiser 1950), and April to July in Czechoslovakia (Vaňhara 1970). Adults are slightly larger than those of *S. lutaria*; e.g. the means and ranges for body length in adults from the Walla Brook and Wilfin Beck were 12 mm (8–15 mm) for males and 16 mm (12–20 mm) for females. Kaiser (1961) observed females apparently looking for food on flowers of cow parsley *Anthriscus silvestris* (L.) Hoffm. Mating has been described by Killington (1932) and is very similar to that of *S. lutaria* except that the male holds the abdomen of the female with one or both forelegs. Rupprecht (1975) has found that males and unmated females generate vibration-signals similar to those produced by *S. lutaria*, except that the vibrations are generated by tapping the abdomen and wings on the ground. Kaiser (1950, 1961, 1977) notes that the eggs are never laid on vegetation along the banks of the stream and are usually found about 3–5 m above the surface of the water on the lower surface of leaves of deciduous trees. The eggs are light brown but are otherwise similar to those of *S. lutaria*. They are usually laid in a circular mass containing between 300 and 1100 eggs. Each egg mass is laid by one female and the smaller egg masses (300–400 eggs) are probably from females that have oviposited for a second time.

Elliott (1995) provides the most detailed information on egg hatching over

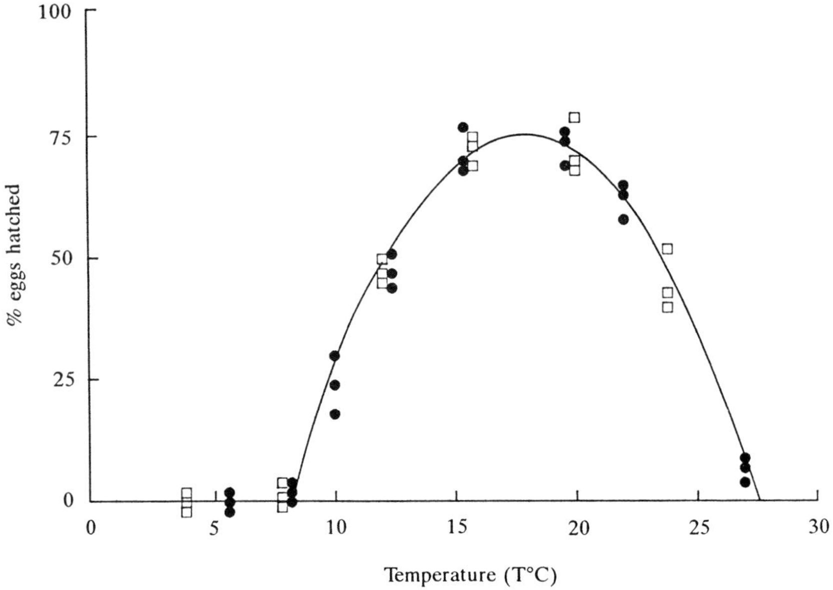

Fig. 22. Relationship between temperature (T °C) and percentage of eggs hatching for *Sialis fuliginosa* from Wilfin Beck (●), and the Walla Brook (□).

a wide range of temperatures. The eggs incubated at ten temperatures came from the Wilfin Beck and Walla Brook populations. The percentage of eggs hatching varied considerably with temperature but there were no significant differences between eggs from the two populations (Fig. 22). No eggs hatched at the two lowest temperatures. The relationship between hatching success and temperature was well described by a second-degree polynomial and 98% of the variation in hatching success could be attributed to temperature variations. This relationship differed from that for *S. lutaria* (Fig. 17) in that the optimum temperature for hatching was lower at 18°C and the range for at least 50% hatched was narrower at 12.3–23.6°C. The percentage of eggs hatching at temperatures close to the optimum was also lower (<80%) than that for *S. lutaria* (>80%). The relationship between the incubation period (d days between oviposition and the date on which 50% of the eggs hatched) and temperature (T °C) was linear on logarithmic scales and was well described by a negative power function (Fig. 18B). There were no significant differences between eggs from the two populations, nor between the relationship for this species and that for *S. lutaria*. It is obvious from Figs 17 and 22 that high temperatures above 25°C would favour hatching in *S. lutaria* but not in *S. fuliginosa*, whereas the opposite would occur at temperatures below 12°C. Such differences between species are important in relation to changes in environmental temperatures due to natural causes or human activities.

There are ten larval instars and the 1st-instar larvae are similar to those of *S. lutaria*. Fully-grown larvae are larger than those of *S. lutaria*; e.g. body length including the terminal filament is usually between 18 and 26 mm in male larvae and 23 and 32 mm in female larvae. Most larvae are in the 6th, 7th or 8th instar by the end of their first year (May or June) and in the 10th instar in autumn (September or October). They finally leave the water for pupation in late April, May or June. Hildrew & Townsend (1976) found that the larvae took prey items roughly in proportion to their numerical abundance in the benthos, the chief food organisms being chironomid larvae and plecopteran nymphs. They later described resource partitioning between this species and a co-existing predator, the net-spinning caddis *Plectrocnemia conspersa* (Curtis) (Townsend & Hildrew 1979; Hildrew & Townsend 1982). Larvae of *S. fuliginosa* are more active than those of *S. lutaria* and often hunt their prey. They are themselves eaten by brown trout (Elliott 1967b).

Pupation and the final emergence of the adult are the same as in *S. lutaria*, except that the pupal stage requires a slightly higher total of 84.5 degree-days above a base temperature of 7.2°C (Kaiser 1961); e.g. it lasts 30 days at 10°C, 20 days at 11.4°C, 10 days at 15.7°C and 5 days at 24.1°C.

Sialis nigripes

Less is known about this species than the two previous species. It appears to be the rarest of the three species in Europe but has now been found in Denmark, Germany, Austria, Spain, Czechoslovakia (references in Vaňhara 1970), England (Barnard 1977; Elliott, O'Connor & O'Connor 1979; Mackey & Ham 1980), Scotland (Fozzard & Clelland 1981), Holland (Mol 1982), France (Leraut 1982), Belgium (Stroot 1986), Ireland (O'Connor & O'Grady 1990). The larvae have been found in large rivers (Elliott, O'Connor & O'Connor 1979; Fozzard & Clelland 1981), fast-flowing small streams (Mackey & Ham 1980; Mol 1982), and lakes (O'Connor & O'Grady 1990). At one of the lake sites, larvae were found in a bed of the bulrush *Scirpus lacustris* L., situated some distance from the shore.

Kaiser (1961) notes that this species "has a very special substrate for the placing of eggs", nearly always using dead vegetable material. Dead plant stems of *S. lacustris* extending above the water surface would therefore form an ideal surface for oviposition. Eggs have also been laid on dead stalks of reed-grass *Phalaris arundinacea* L. (Fozzard & Clelland 1981). The eggs are usually found 0.5–3 m above the water surface, and have a long micropylar knob with a length of 0.3–0.4 mm, compared with about 0.1 mm in the other two species (Kaiser 1961, 1977). The egg mass is often square and contains 800–900 eggs. Some eggs have been found with a round emergence hole of the type made by the parasitoid hymenopteron *Trichogramma semblidis*, which also attacks *S. lutaria* (Barnard 1977).

From the limited information, the flight period appears to be May and June in Britain and Ireland, and this agrees with observations from Denmark (Kaiser 1950), Czechoslovakia (Vaňhara 1970), and Belgium (Stroot 1986). Kaiser (1961) notes that adults of this species, unlike the other two *Sialis* species, are often found in isolated clumps of trees or bushes some distance from water. Length ranges for larvae (without terminal filament) from Denmark were 14–17 mm for males and 16–21 mm for females (Kaiser 1961). Pupation and adult emergence are the same as in the other two species, and the pupal stage requires a total of 82.4 degree-days above a base temperature of 7.8°C, the latter being the highest for the three *Sialis* species (Kaiser 1961); e.g. it lasts 30 days at 10.6°C, 20 days at 11.9°C, 10 days at 16.0°C and 5 days at 24.3°C.

Very little is therefore known about the life history and habitat requirements of this species. Its distribution in Ireland indicates that it may be associated chiefly with calcareous rivers, streams and lakes (O'Connor & O'Grady 1990).

Genus *OSMYLUS*

The family Osmylidae has a wide distribution, occurring in Europe, Asia, Africa, New Zealand, Australia and South America. Two species, both in the genus *Osmylus*, occur in Europe but one species, *O. elegantissimus* K., is restricted to the Caucasian region whilst the other, *O. fulvicephalus* (Scop.), occurs throughout most of Europe, including Britain and Ireland (Aspöck, Aspöck & Hölzel 1978; Barnard 1978; Barnard, O'Connor & Speight 1991).

Osmylus fulvicephalus

This is a local species but usually forms colonies so that it is fairly abundant where it is found. Adults are often found under bridges and near woodland streams with a dense cover of overhanging vegetation on their banks. Larvae live in damp moss at the edge of the stream or on overhanging banks. Early observations on the life cycle were made by Stein (1838), Brauer (1850, 1851), Hagen (1852) and McLachlan (1868), but more detailed accounts are provided by Withycombe (1923), David (1936), Killington (1936) and Ward (1965).

The adults are the largest of the British Neuroptera (Fig. 3C). Their flight period is usually from May to July but they have occasionally been taken in April and August (Plant 1994). The length of life of a male is only a few weeks but females can live for as long as two to three months (Withycombe 1923). They are feeble flyers and are most active at dusk when they feed on small insects including aphids (Withycombe 1923) and adult Lepidoptera which are probably caught when resting, injured or even dead (Killington 1936). A detailed study of feeding in adults of forty-nine species of Neuroptera has shown that *Osmylus fulvicephalus* feeds not only on insects but also on honeydew to which is stuck various particles such as pollen grains, fungal spores and algae (Stelzl 1991, 1992).

Courtship is unusual because the male is the more passive sex and attracts the female. He has a pair of whitish, finger-shaped, eversible scent-glands between the 8th and 9th abdominal segments and these can be extruded to a length of about 4 mm. Withycombe (1923) first described the courtship and mating in detail. When the male is ready for mating, he hangs down from a support, lifts his wings and everts his scent-glands. Females in the immediate vicinity respond almost at once by waving their antennae vigorously, then flying or walking towards the male and stroking the scent-glands with their antennae and palps. The male withdraws the scent-glands and turns to meet the female. After male and female have stroked their antennae for a few minutes, the male carefully bites on one of the female's fore-coxae, each of which carries a teat-like process. Both insects bend their abdomens to meet, the apex of the female's abdomen curling over and then under that of the male so that the elongate ventral valves of the female grasp the genitalia of the male. The male now releases his hold on the female and the ventral valves of the female lever on the male genitalia for between ten and sixty minutes until a large white spermatophore is produced by the male. After the separation of the two insects, the spermatophore is seen projecting from the apex of the female's abdomen. The spermatophore is about 4 mm long, white in colour but yellowish at the centre, and consists of four rounded lobes and a short attaching stem. Soon after the female has separated from the male, she bends her head under her body and eats the nearest lobe of the spermatophore. If she continues to eat the spermatophore, the male intervenes and caresses her. When the female finally leaves the male, the remaining part of the spermatophore is either eaten or becomes dislodged. A female may mate two or three times but once is sufficient for all the eggs laid to be fertile.

The life cycle takes one year and has been studied in the Cunsey Beck in the English Lake District (Fig. 23A). Eggs are laid 2-3 days after mating in late May, June and early July. Each female lays ca. 30 eggs in small batches on leaves, twigs and especially the leaves of moss near the water (Lestage 1920b; David 1936; Ward 1965). The eggs are cylindrical, slightly flattened,

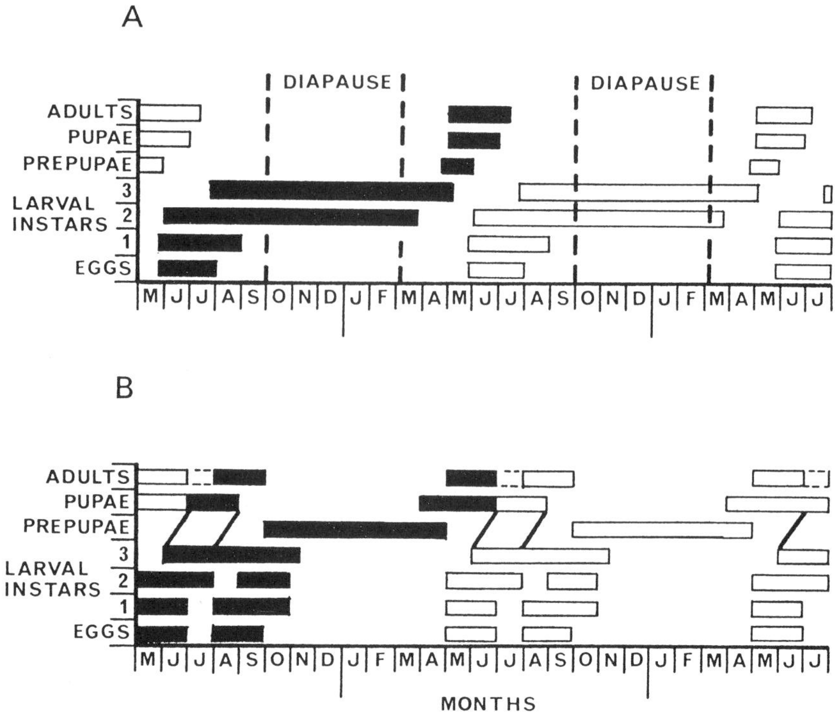

Fig. 23. Life cycle of: A, *Osmylus fulvicephalus*; B, *Sisyra fuscata*; both from the Cunsey Beck in the English Lake District. Each horizontal bar indicates the months in which the various stages of the life cycle occur.

with a short micropylar knob, a length of between 1.6 and 1.8 mm and a breadth of between 0.5 and 0.6 mm. They are glossy-white when laid but darken to creamy-yellow or brown within a few days. Eggs deposited on moss are laid singly or in pairs on the underside of leaves with the micropyle towards the tip of the leaf. They are also laid in regular, straight or slightly curved rows on larger objects, each row with 2–20 eggs touching one another. The eggs hatch within 4–22 days, depending upon the temperature (Withycombe 1923; David 1936; Ward 1965), and the larva escapes through a slit made by a very long, saw-like egg-breaker. Within one to three days after emergence, the larvae leave the locality of the eggs and burrow into the moss.

There are only three larval instars. The 1st instar is slightly different from the other two instars; the head is broader than the thorax, and the jaws, setae and legs are relatively longer than those of later instars, and the fore-gut contains one or two bubbles of air which may prevent the larva from sinking under water (Lestage 1920b; Withycombe 1923). 2nd and 3rd instars are very similar (Fig. 10A, p.25). Ward (1965) measured large numbers of captive larvae at intervals of one to three days and obtained the following mean values and ranges: overall length (inclusive of jaws), 1st instar 4.95 mm (3.53–5.93 mm), 2nd instar 8.91 mm (7.14–10.39 mm), 3rd instar 15.70 mm (13.50–16.20 mm); width of head capsule, 1st instar 0.54 mm (0.50–0.59 mm), 2nd instar 0.86 mm (0.80–0.88 mm), 3rd instar 1.54 mm (1.53–1.56 mm).

There is some disagreement over whether the larvae are truly aquatic, semi-aquatic or terrestrial. Several workers have stated that the larvae are definitely aquatic (e.g. Brauer 1850, 1851; Girard 1879; Withycombe 1923), but Hagen (1852) and Brocher (1913) considered that they were terrestrial. Other authors have found that the larvae live in damp moss and debris near the water but may enter the water in search of prey (McLachlan 1868; David 1936; Killington 1936). Ward (1965) found that when larvae were dropped into water, they rapidly crawled out, and that when larvae inside moss were immersed in water, they crawled deep into the moss and died within 8–28 days. He therefore concluded that although the larvae live near the water's edge, they never enter the water of their own accord and cannot be regarded as aquatic or even semi-aquatic.

The larvae are carnivores and 1st instars feed chiefly on small arthropods, such as mites and Collembola, whereas 2nd and 3rd instars feed chiefly on dipterous larvae, such as Chironomidae and Tipulidae. If a movement is detected in the moss or mud on which the larva is walking, the larva stabs downwards with its long, slender jaws. When the tips of the jaws strike a prey, a salivary secretion is injected into the prey which is quickly paralysed. The contents of the prey are then slowly removed by the sucking jaws. The paralysing effect of the secretion is far more rapid in chironomid larvae than in surface-dwelling arthropods such as aphids and mites, and a large chironomid larva is paralysed after only ten seconds (Withycombe 1923).

The growth rate of the larvae depends upon the food supply and temperature. Ward (1965) found that the duration of each instar in captivity was betwen 14 and 24 days for 1st instars, 28 and 39 days for 2nd instars, and 248 days for a 3rd-instar larva. All larvae are in the 2nd or 3rd instar by the end of the summer. In late September or October, the larvae cease to feed, and burrow deep into the roots of the moss where they overwinter in a state of hibernation known as diapause. Larvae in this condition can survive total immersion in water and this may happen frequently during winter spates (David 1936). Hibernation ends in March and the larvae start to feed again.

In late April or early May, a silken cocoon is spun amongst the moss, some moss being incorporated into the cocoon. The cocoon is a thin, closely woven, yellowish-white structure, irregularly oval in shape, about 10 mm long and 8 mm broad. The larva remains in a prepupal stage in the cocoon for 7–18 days and then pupates. Just prior to pupation, the long jaws of the larva break off near their base. The pupal appendages are free but immovable until just before emergence. There is a pair of well-developed pupal mandibles and the head and abdomen are bent ventrally. The pupal stage lasts ca. 10–14 days and then the pupa cuts an irregular slit in the cocoon with its mandibles. After the pupa has crawled to some firm support, the final moult occurs, usually in the morning or evening, and the adult emerges.

A summary of the one-year life cycle is shown in Fig. 23A. As the life cycle will vary from one locality to another, this figure can only serve as a guide for other populations.

Genus *SISYRA*

The family Sisyridae has a wide distribution, occurring in Europe, Asia, Africa, North America, South America, the West Indies, the Philippine Islands and Australia. Taxonomic works on the whole family include monographs by Krüger (1923), Navás (1935) (reviewed by Lestage 1935), and Parfin & Gurney (1956). The family appears to be closely related to the Osmylidae in spite of the very different appearance of the imagines. Withycombe (1925) considered that the Sisyridae were derived from osmyloid ancestors, the larvae of which took to deep water and then to preying on freshwater sponges. He suggested that the larvae developed abdominal tracheal gills, and longer and slimmer jaws, that their labial palpi were lost, and that the two claws on each tarsus fused laterally into a single claw.

The family Sisyridae is represented world-wide by only about fifty species in three genera. Twenty-three species have now been recorded in the genus *Sisyra*. Only five species occur in Europe and three of these have been found in Britain. Several workers have studied the life cycle and general biology of *Sisyra fuscata* but, until the recent study by Weissmair (1994), little was known about *S. terminalis*. There is still a paucity of information on *S. dalii*.

Sisyra fuscata

This species is widely distributed in Europe and is the only sisyrid known to occur in both the Nearctic and Palaearctic regions (Aspöck, Aspöck & Hölzel 1978). Early accounts of some aspects of the ecology of this species were provided by Withycombe (1923, 1925) and Killington (1936), and there is a recent detailed account by Weissmair (1994). In Britain and Ireland, the flight period is from May to September, very occasionally October, but adults are most abundant in June, July and August (Plant 1994). A similar flight period has been recorded in Austria (Weissmair 1994). Adults were taken in a light trap in Norway from May to early September and 70% of the catch were males (Andersen & Greve 1975). Larvae are found both on and inside sponges and bryozoans inhabiting ponds, lakes, canals, rivers and streams, and the adults are usually found on the vegetation near the water.

The life cycle takes one year or less, and in some localities there may be two broods in one year. The following account is based chiefly on a study of a population in the Cunsey Beck in the English Lake District (Fig. 23B). Eggs were laid in May and June and most of the larvae from these eggs grew slowly through the summer, left the water in autumn, overwintered in the prepupal stage within the cocoon, and pupated in April, May and June. The adults finally emerged in May and June to complete a one-year cycle. A small number of larvae from the eggs laid in May and June grew rapidly to reach the pupal stage in a few weeks and the adult stage in August and September to complete the life cycle in 4–5 months. These adults laid eggs in August and September, the larvae grew rapidly, overwintered in the prepupal stage and pupated in April, May and June. This complex life cycle will vary from one locality to another, and there may be no rapidly-growing second brood in some localities. However, a life cycle similar to that depicted in Fig. 23B has been described for an Austrian population with adults present from mid-May to mid-July (first generation) and from mid-July to mid-September (second generation) (Weissmair 1994).

The adults are most active in the evening and sometimes form small swarms over the water. Tjeder (1944) observed an adult male feeding on eggs of *Sialis lutaria*, and other sisyrids feed upon plant products such as nectar (Brown 1952). A detailed study of feeding in adults of forty-nine species of Neuroptera has shown that *Sisyra fuscata* feeds not only on insects and mites but also on honeydew to which is stuck various particles such as pollen grains, fungal spores and algae (Stelzl 1991). The mouthparts have spoon-like structures (lacinia) that are an adaptation for the consumption of honeydew, and they also have grooves for catching mites on the maxillary palps which act as a comb (Stelzl 1992). Adults are eaten by birds, spiders and predaceous insects, and Berg (1948) found them infested with larvae of the mite family Trombidiidae.

Mating occurs at dusk and several hours to two weeks after emergence. There is no elaborate courtship and the male copulates laterally with the female. After about 3–5 minutes, a small white spermatophore is left attached to the tip of the abdomen of the female and is immediately eaten. Oviposition takes place in the evening, often only a few hours after pairing, and the eggs are laid in small clusters of 1–12 eggs. The blade-shaped 9th sternite (Fig. 7E) contains the opening of the cement gland and is also used for locating crevices in which to deposit eggs on leaves, wooden piles and other objects overhanging the water. Each batch of eggs is covered by a web of white silk. The female spins a few parallel strands, then shifts her position slightly and spins a few more strands, crossing the first strands, and so on until three or four layers of silk are laid down. The eggs are pale yellow, elongate oval, with a small, flattened micropylar knob and a length of about 0.35 mm.

Hatching usually occurs within 8–14 days, the incubation period being partially dependent upon the temperature. The larva uses its egg-breaker to saw a slit in the egg shell, works its way out of the shell, pushes through the layer of silk and falls onto the surface of the water. It then penetrates the surface film by bending the tip of the abdomen back over the dorsal side of the head, so that the anterior part of the body is forced under the water as the body is suddenly straightened. The submerged larva floats in the water, occasionally swims, and drifts about until it comes into contact with a sponge or bryozoan. Weissmair & Waringer (1994) note that the 1st-instar larvae drifted for 24–168 hours, depending upon the water temperature. Once it attaches to a sponge, the larva starts to feed by inserting its jaws and sucking the fluids of the sponge. The usual hosts are *Spongilla lacustris* (L.) and *Ephydatia fluviatilis* (L.), but Weissmair & Waringer (1994) list other sponge and bryozoan hosts.

There are only three larval instars. The 1st-instar larva is slightly different from those of the other two instars; the jaws are stouter and shorter, the abdomen is narrower than the thorax, there are very long setae on the 9th abdominal segment and abdominal gills are absent (similar to that of *S. terminalis* in Fig. 13A, p.33). Body lengths of the three instars are given in the keys on pp. 32 and 34.

Fully-fed larvae in the 3rd instar swim away from the sponge to the shore. Whilst swimming, they are rather conspicuous and must be an easy prey for fish. On leaving the water, they frequently travel a considerable distance before selecting a site for spinning a cocoon, and are therefore an obvious prey for predatory arthropods and birds. Cocoons are sometimes found in great numbers on the walls of bridges and in crevices of bark or under loose bark on tree trunks. The cocoon has a double structure; an outer, coarse, yellowish silk network and an inner, finer, white cocoon. When placed in a crevice, the outer envelope forms a canopy over the inner, the whole structure being about 4 mm by 3.5 mm. The larvae that leave the water in autumn spend the winter as resting larvae (prepupae) in the pupal cocoon and finally pupate in early spring. Larvae leaving the water in mid-summer pupate immediately. Large numbers of prepupae and a few pupae are sometimes killed by a white mould (Killington 1936), and a parasitic hymenopteron,

Eupteromalus sp., which lays an egg beside the resting larva (Killington 1933). The pupal and prepupal stages can be also parasitised by the hymenopteron *Gyrinophagus aper* (Walker) (Weissmair 1994).

The pupa is short and stout with large mandibles and free appendages. After a short pupal period of about two weeks, an irregular hole is cut through the anterior end of the cocoon by the pupal mandibles. The pupa does not always leave the cocoon before the emergence of the adult. Most adults emerge in the evening, after sunset.

Sisyra terminalis

This species occurs in western and central Europe, but appears to be absent from northern Scandinavia and the extreme southeast of Europe (Aspöck, Aspöck & Hölzel 1978). From distribution records, Plant (1994) concluded that it is the rarest of the three *Sisyra* species found in Britain and Ireland, being taken chiefly in southern Ireland, near Bangor in Wales, the west Midlands and southeast England. Most records were for adults taken in light traps, but on one occasion, several hundred adults were seen swarming around bushes overhanging the River Thames. Withycombe (1923) claimed to have found all stages of this species and states that they are hardly distinguishable from those of *S. fuscata*. The keys provided earlier (pp. 32 and 34) show that it is now possible to separate the two species.

Sisyra terminalis appears to prefer water overhung with the branches of oak trees, so it is possible that there is an association with oaks which perhaps serve as sites for oviposition or overwintering, or as a source of insect food (Barnard, O'Connor & Speight 1991; Plant 1994). The gut contents of the adults shows that they feed on mites and insects, chiefly aphids (Kokobu & Duelli 1982; Eglin-Dederding 1985). The capture-records for adults suggest that the larvae occur chiefly in large rivers, occasionally in lakes.

The flight period in Britain is from late May to late August, but nothing is known about the life cycle. The only detailed study of the latter is for an Austrian population from the flowing backwaters within the riverine forests of the Danube (Weissmair 1994; Weissmair & Waringer 1994). As in *S. fuscata*, there were two generations and these were on the wing from mid-June to mid-July and from August to September. However, unlike *S. fuscata*, one generation of the Austrian population of *S. terminalis* overwintered in the 1st and 2nd larval instars as well as in the prepupal stage. The flight period for *S. terminalis* also commenced about one month after that of *S. fuscata*. It is possible that there is only one generation of *S. terminalis* at some localities. For example, Langford (1975) caught adults of this species in emergence traps floating on the River Severn in England and found that the emergence period was very short, being restricted to late June and early July. He concluded that this species was univoltine at this site.

Sisyra dalii

The European distribution of this species appears to be more restricted than those of the other two *Sisyra* species, with positive records from only the Pyrenees, east Balkans, west and central Mediterranean, Denmark, south Sweden, Norway, as well as Britain and Ireland (Aspöck, Aspöck & Hölzel 1978). There is a single recent Irish record from west Galway and older records from Dublin, Kerry and Wexford (Barnard, O'Connor & Speight 1991; Plant 1994). Most of the British records are from Wales, Shropshire, the English Lake District, and the extreme south of England from Lands End to near London (Plant 1994). These records for adults suggest that the chief habitat of the larvae is a fast-flowing upland river, but there are also captures of adults near small lakes and canals. The flight period is from early June to late August (Plant 1994). No one has yet found the larvae and virtually nothing is known about the ecology of this species.

ACKNOWLEDGEMENTS

I wish to express my sincere thanks to the following for their help in the preparation of this book: Dr D. W. Sutcliffe for editing the manuscript and supervising its conversion to the final publication; Dr R. Sweeting for facilitating the support of the National Rivers Authority; T. I. Furnass for his assistance in the preparation of the illustrations and for designing the front cover of this publication; Mrs Joan Worthington for her excellent illustrations; E. W. Kaiser of Klokkedal, Denmark, for many helpful suggestions and for sharing his extensive knowledge of the genus *Sialis*; W. Weissmair for his help with the genus *Sisyra*; Dr P. C. Barnard of the British Museum (Natural History) for helpful advice and for the loan of specimens; Dr P. D. Armitage, Professor A. G. Hildrew, R. A. Jenkins, Dr M. A. Learner, Mrs M. J. Morgan and Dr R. S. Wilson for supplying specimens.

REFERENCES

Andersen, T. & Greve, L. (1975). Neuroptera in light-traps at Osterøy, Hordaland. *Norw. J. Ent.*, **22**, 123–128.

Aspöck, H., Aspöck, U. & Hölzel, H. (1978). Megaloptera et Plannipennia. In *Limnofauna Europaea* (ed. J. Illies), 329–332. Stuttgart. Fischer.

Aurivillius, C. (1898). En ny Svensk äggparasit. *Ent. Tidskr.*, **18**, 249–56.

Azam, K. M. & Anderson, N. H. (1969). Life history and habits of *Sialis rotunda* and *S. californica* in Western Oregon. *Ann. ent. Soc. Am.*, **62**, 549–558.

Barnard, P. C. (1977). *Sialis nigripes* Pictet (Megaloptera, Sialidae), an alder–fly new to Britain. *Entomologist's Gaz.*, **28**, 269–274.

Barnard, P. C. (1978). A check-list of the British Neuroptera with taxonomic notes. *Entomologist's Gaz.*, **29**, 165–176.

Barnard, P. C., O'Connor, J. P. & Speight, M. C. D. (1991). A review of published data for Irish Neuroptera (Insecta), together with additional records and a check-list of the Irish species. *Bull. Ir. biogeog. Soc.*, **14**, 109–122.

Batson, B. S. (1985). A paradigm for the study of insect-ciliate relationships: *Tetrahymena sialidos* sp. nov. (Hymenostomatida: Tetrahymenidae), parasite of larval *Sialis lutaria* (Linn.) (Megaloptera: Sialidae). *Phil. Trans. R. Soc. Lond.*, B, **310**, 123–144.

Beadle, L. C. & Shaw, J. (1950). The retention of salt and the regulation of the non-protein nitrogen fraction in the blood of the aquatic larva, *Sialis lutaria*. *J. exp. Biol.*, **27**, 96–109.

Berg, K. (1938). Studies on the bottom animals of Esrom Lake. *K. danske Vidensk. Selsk. Skr. (naturv. math.)*, ser. 9, **8**, 1–255.

Berg, K. (1948). Biological studies on the River Susaa. *Folia limnol. scand.*, **4**, 1–318.

Berg, K. & Petersen, I. B. (1956). Studies on the humic acid Lake Gribsø. *Folia limnol. scand.*, **8**, 1–273.

Bertrand, H. (1949). Notes morphologiques sur les larves des *Sialis* L. (Mégaloptères, Sialidae). *Feuille Nat.*, **4**, 5–10.

Brauer, F. (1850). Ueber die Verwandlung des *Osmylus maculatus*. *Ber. Mitt. Freunden Naturw. Wien*, **7**, 153.

Brauer, F. (1851). Verwandlungsgeschichte des *Osmylus maculatus*. *Arch. Naturgesch.*, **17**, 255–258.

Brocher, F. (1913). *L'Aquarium de Chambre*. Lausanne, 317 pp.

Brooker, M. P. (1979). The life cycle and growth of *Sialis lutaria* L. (Megaloptera) in a drainage channel under different methods of plant management. *Ecol. Ent.*, **4**, 111–117.

Brown, H. P. (1952). The life history of *Climacia areolaris* (Hagen), a neuropterous 'parasite' of freshwater sponges. *Am. Midl. Nat.*, **47**, 130–160.

Dall, P. C. (1987). Larvemigration hos *Sialis lutaria* L. (Megaloptera) i Esrom Sø . *Flora og Fauna*, **93**, 47–50.

Dall, P. C. (1989). The life history and occurrence of *Sialis lutaria* L. (Megaloptera) in the littoral zone of Lake Esrom, Denmark. *Int. Revue ges. Hydrobiol.*, **74**, 273–281.

David, K. (1936). Beiträge zur Anatomie und Lebensgeschichte von *Osmylus chrysops* L. *Z. Morph. Ökol. Tiere*, **31**, 151–206.

Du Bois, A. M. (1936). Recherches expérimentales sur la détermination de l'embryon dans l'oeuf de *Sialis lutaria. Revue suisse Zool.*, **43**, 519–523.

Du Bois, A. M. & Geigy, R. (1935). Beiträge zur Ökologie, Fortpflanzungbiologie und Metamorphose von *Sialis lutaria L. Revue suisse Zool.*, **42**, 169–248.

Eglin-Dederding, W. (1985). Die Süsswasserschwammfliege *Sisyra terminalis* Curtis (Neuropteroidea: Plannipennia: Sisyridae) immer noch häufig im Rhein bei Basel. *Mitt. Entom. Ges. Basel*, **35**, 66–68.

Elliott, J. M. (1967a). Invertebrate drift in a Dartmoor stream. *Arch. Hydrobiol.*, **63**, 202–237.

Elliott, J. M. (1967b). The food of trout (*Salmo trutta*) in a Dartmoor stream. *J. appl. Ecol.*, **4**, 59–71.

Elliott, J. M. (1973). The life cycle and production of the leech *Erpobdella octoculata* (L.) (Hirudinea: Erpobdellidae) in a Lake District stream. *J. Anim. Ecol.*, **42**, 435–448.

Elliott, J. M. (1977). A key to the larvae and adults of British freshwater Megaloptera and Neuroptera with notes on their life cycles and ecology. *Scient. Publs Freshwat. Biol. Ass.* No. **35**, 52 pp.

Elliott, J. M. (1995). The effect of temperature on egg hatching for three populations of *Sialis lutaria* (L.) and two populations of *Sialis fuliginosa* Pictet (Megaloptera: Sialidae). *Entomologist's Gaz.*, **46**, 155–159.

Elliott, J. M., O'Connor, J. P. & O'Connor, M. A. (1979). A key to the larvae of Sialidae (Insecta: Megaloptera) occurring in the British Isles. *Freshwat. Biol.* **9**, 511–514.

Elliott, J. M., Tullett, P. A. & Elliott, J. A. (1993). A new bibliography of samplers for freshwater benthic invertebrates. *Occ. Publs Freshwat. Biol. Ass.*, No. **30**, 92 pp.

Fozzard, I. & Clelland, B. E. (1981). The occurrence of *Sialis nigripes* Pictet (Megaloptera: Sialidae) in central Scotland. *Entomologist's Gaz.*, **32**, 268–270.

Fraser, F. C. (1959). Mecoptera, Megaloptera and Neuroptera. *Handbk Ident. Br. Insects*, **1**, 12–13, 40 pp.

Geigy, R. (1937). Beobachtungen über die Metamorphose von *Sialis lutaria* L. *Mitt. schweiz. ent. Ges.*, **17**, 144–157.

Geigy, R. (1948). Étude expérimentale de la métamorphose de *Sialis lutaria* L. *Bull. biol. Fr. Belg.*, **55**, 62–67.

Geigy, R. & Du Bois A. M. (1935). Sinnesphysiologische Beobachtungen über die Begattung von *Sialis lutaria* L. *Revue suisse Zool.*, **42**, 447–457.

Geigy, R. & Grobe, D. (1958). Die ökologische Abhängigkeit des Metamorphose-Geschehens bei *Sialis lutaria* L. *Revue suisse Zool.*, **65**, 323–328.

Geigy, R. & Ochsé, W. (1940a). Schnürungsversuche an Larven von *Sialis lutaria* L. *Revue suisse Zool.*, **47**, 193–194.

Geigy, R. & Ochsé, W. (1940b). Versuche über die inneren Faktoren der Verpuppung bei *Sialis lutaria* L. *Bull. biol. Fr. Belg.*, **47**, 225–241.

Geigy, R. & Rahm, U. H. (1951). Beiträge zur experimentellen Analyse der Metamorphose von *Sialis lutaria* L. *Bull. biol. Fr. Belg.*, **58**, 408–413.

Giani, N. & Laville, H. (1973). Cycle biologique et production de *Sialis lutaria* L. (Megaloptera) dans le lac de Port-Bielh (Pyrénées Centrales). *Annls Limnol.*, **9**, 45–61.

Girard, M. (1879). *Traité Elementaire d'Entomologie*, **2**, 443–456.

Griffiths, D. (1973). The food of animals in an acid moorland pond. *J. Anim. Ecol.*, **42**, 285–293.

Hagen, H. A. (1852). Die Entwicklung und der innere Bau von *Osmylus*. *Linnaea Ent., Berl.*, **7**, 368–418.

Hildrew, A. G. & Townsend, C. R. (1976). The distribution of two predators and their prey in an iron rich stream. *J. Anim. Ecol.*, **45**, 41–57.

Hildrew, A. G. & Townsend, C. R. (1982). Predators and prey in a patchy environment: a freshwater study. *J. Anim. Ecol.*, **51**, 797–815.

Hoffmann, J. (1962). Faune des Neuroptéroides du Grand-Duché de Luxembourg. *Archs Inst. gr.-duc. Luxemb.*, **28**, 249–332.

Humphries, C. F. (1936). An investigation of the profundal and sublittoral fauna of Windermere. *J. Anim. Ecol.*, **5**, 29–52.

Iversen, T. M. & Thorup, J. (1987). Population dynamics and production of *Sialis lutaria* L. (Megaloptera) in the Danish River Suså. *Freshwat. Biol.* **17**, 461–469.

Jacques, D. (1974). *The Development of Modern Stillwater Fishing.* London. A. & C. Black, 237 pp.

Kaiser, E. W. (1950). *Sialis nigripes* Ed. Pict., new to Denmark, and the distribution of *S. lutaria* and *S. fuliginosa* Pict. in Denmark. *Flora Fauna, Århus*, **56**, 17–36.

Kaiser, E. W. (1961). On the biology of *Sialis fuliginosa* Pict. and *S. nigripes* Ed. Pict. *Flora Fauna, Århus*, **67**, 74–96.

Kaiser, E. W. (1977). Eggs and larvae of six *Sialis*–species from Scandinavia and Finland (Megaloptera, Sialidae). *Flora Fauna, Århus*, **83**, 65–79.

Kennedy, C. R., Nie, P. & Rostron, J. (1992). An insect, *Sialis lutaria*, as a host for larval *Proteocephalus* sp. *J. Helminthol.*, **66**, 7–16.

Killington, F. J. (1926). Notes on Neuroptera taken in 1925. *Entomologist*, **59**, 110–112.

Killington, F. J. (1929). A synopsis of British Neuroptera. *Trans. ent. Soc. S. Engl.*, **5**, 1–36.

Killington, F. J. (1932). On the pairing of *Sialis fuliginosa* Pict. (Neuroptera : Megaloptera). *Entomologist*, **65**, 66–67.

Killington, F. J. (1933). The parasites of Neuroptera with special reference to those attacking British species. *Trans. ent. Soc. S. Engl.*, **8**, 84–91.

Killington, F. J. (1936). *A Monograph of the British Neuroptera*, **1**. London. Ray Society. 269 pp.

Killington, F. J. (1937). *A Monograph of the British Neuroptera*, **2**. London. Ray Society. 306 pp.

Kimmins, D. E. (1962). Keys to the British species of aquatic Megaloptera and Neuroptera with ecological notes. *Scient. Publs Freshwat. Biol. Ass.* No. **8**, 23 pp.

Kokobu, H. & Duelli, P. (1982). Adult food of sponge flies: observations on the crop and gut content of *Sisyra terminalis* Curtis (Neuropt. Plannipennia: Sisyridae). *Neuroptera International*, **II**, 157–162.

Krüger, L. (1923). Sisyridae. Beiträge zu einer Monographie der Neuropteren-Familie der Sisyriden. *Stettin. ent. Ztg*, **84**, 25–66.

Kryger, J. P. (1919). The European Trichogramminae. *Ent. Meddr*, **12**, 257–354.

Langford, T. E. (1975). The emergence of insects from a British river, warmed by power station cooling-water. *Hydrobiologia*, **47**, 91–133.

Lassiere, O. L. (1988). Host-parasite relationships between larval *Sialis lutaria* (Megaloptera) and *Neoechinorhynchus rutili* (Acanthocephala). *Parasitology*, **97**, 331–338.

Leraut, P. (1982). *Sialis nigripes*, espece nouvelle pour la France (Megaloptera, Sialidae). *Entomologiste*, **38**, 145.

Lestage, J. A. (1919). Notes biologiques sur *Sialis lutaria* L. (Megaloptera). *Annls Biol. lacustre*, **9**, 25–38.

Lestage, J. A. (1920a). Le méchanisme de la ponte chez *Sialis lutaria* L. (Megaloptera). *Annls Biol. lacustre*, **10**, 221–223.

Lestage, J. A. (1920b). La ponte et la larvule de l'*Osmylus chrysops* L. (Planipenne). *Annls Biol. lacustre*, **10**, 226–230.

Lestage, J. A. (1935). Notes sur les sisyridés (Hémérobiiformes à larve aquatique). *Bull. Annls Soc. r. ent. Belg.*, **75**, 387–394.

Lundbeck, J. (1926). Die Bodentierwelt norddeutscher Seen. *Arch. Hydrobiol. (Suppl.)*, **7**, 1–473.

Macan, T. T. (1966). The influence of predation on the fauna of a moorland fishpond. *Arch. Hydrobiol.*, **61**, 432–452.

Macan, T. T. & Worthington, E. B. (1972). *Life in Lakes and Rivers.* (Revised paperback edition). London. Fontana. 320 pp.

Mackey, A. P. & Ham, S. F. (1980). New records of *Sialis* (Megaloptera: Sialidae) from the River Coln, Gloucestershire. *Entomologist's mon. Mag.*, **115**, 46.

Matthey, W. (1971). Écologie des insectes aquatiques d'une tourbière du Haut-Jura. *Revue suisse Zool.*, 78, 367–536.

McLachlan, A. J. & McLachlan, S. M. (1975). The physical environment and bottom fauna of a bog lake. *Arch. Hydrobiol.*, 76, 198–217.

McLachlan, R. (1868). A monograph of the British Neuroptera – Plannipennia. *Trans. Ent. Soc. Lond.*, 1868, 145–224.

Meinander, M. (1962). The Neuroptera and Mecoptera of Eastern Fennoscandia. *Fauna fenn.*, **13**, 1–96.

Miall, L. C. (1895). *The Natural History of Aquatic Insects.* London. Macmillan, 395 pp.

Mol, A. W. M. (1982). *Sialis nigripes* Pictet in Nederland (Megaloptera: Sialidae). *Ent. Ber.*, **42**, 177–179.

Navás, L. (1935). Monografía de la Familia de los sisirídos (Insectos neuropteros). *Mems Acad. Cienc. exact. fís.-quím. nat. Zaragoza*, **4**, 1–86.

Ochsé, W. (1944). Experimentelle und histologische Beiträge zur inneren Metamorphose von *Sialis lutaria* L. *Revue suisse Zool.*, **51**, 1–82.

O'Connor, J. P. & Norton, M. A. (1977). Athlone field meeting: preliminary notes on the aquatic invertebrate fauna of Hare Island and environs. *Bull. Ir. biogeog. Soc.*, **1**, 20–25.

O'Connor, J. P. & Norton, M. A. (1978). Further notes on the aquatic invertebrates of Hare Island and its environs. *Bull. Ir. biogeog. Soc.*, **2**, 37–39.

O'Connor, J. P. & O'Grady, M. F. (1990). Observations on the distribution of *Sialis nigripes* Pictet (Megaloptera) in Ireland. *Entomologist's Gaz.*, **41**, 109–110.

Parfin, S. I. & Gurney, A. B. (1956). The Spongilla-flies, with special reference to those of the Western Hemisphere (Sisyridae, Neuroptera). *Proc. U.S. natn. Mus.*, **105**, 421–529.

Plant, C. W. (1994). *Provisional Atlas of the Lacewings and Allied Insects (Neuroptera, Megaloptera, Raphidioptera and Mecoptera) of Britain and Ireland.* Monks Wood. NERC Institute of Terrestrial Ecology. 208 pp.

Pritchard, G. & Leischner, T. G. (1973). The life history and feeding habits of *Sialis cornuta* Ross in a series of abandoned beaver ponds (Insecta: Megaloptera). *Can. J. Zool.*, **51**, 121–131.

Rahm, V. H. (1952). Die innersekretorische Steuerung der postembryonalen Entwicklung von *Sialis lutaria* L. *Revue suisse Zool.*, **59**, 179–237.

Rupprecht, R. (1975). Die Kommunikation von *Sialis* (Megaloptera) durch Vibrationssignale. *J. Insect Physiol.*, **21**, 305–320.

Rupprecht, R. (1983). A contribution to the fauna of Sialidae in Scandinavia. *Fauna Norrlandica*, **2**, 1–11.

Salt, G. (1937). The egg parasite of *Sialis lutaria*: a study of the influence of the host upon a dimorphic parasite. *Parasitology*, **29**, 539–553.

Seitz, W. (1940). Zur Frage des Extremitütencharakters der Tracheenkiemen von *Sialis flavilatera* L. *Z. Morph. Ökol. Tiere*, **37**, 214–275.

Selman, B. J. (1960a). On the tissue isolated in some of the larval appendages of *Sialis lutaria* L. at the larval-pupal moult. *J. Insect Physiol.*, **4**, 235–257.

Selman, B. J. (1960b). Tolerance of dehydration of the blood of *Sialis lutaria* L. *J. Insect Physiol.*, **6**, 81–83.

Selman, B. J. (1962). The fate of the blood cells during the life history of *Sialis lutaria* L. *J. Insect Physiol.*, **8**, 209–214.

Selman, B. J. (1965). The circulatory system of the alder fly *Sialis lutaria* L. *Proc. zool. Soc. Lond.*, **144**, 487–535.

Shaw, J. (1955a). The permeability and structure of the cuticle of the aquatic larva of *Sialis lutaria*. *J. exp. Biol.*, **32**, 330–352.

Shaw, J. (1955b). Ionic regulation and water balance in the aquatic larva of *Sialis lutaria*. *J. exp. Biol.*, **32**, 353–382.

Solem, J. O. (1973). The bottom fauna of Lake Lille-Jonsvann, Trondelag, Norway. *Norweg. J. Zool.*, **21**, 227–261.

Speight, M. C. D. (1977). Athlone field meeting: insects collected on and near the shores of Lough Ree. *Bull. Ir. biogeog. Soc.*, **1**, 27–36.

Staddon, B. W. (1955). The excretion and storage of ammonia by the aquatic larva of *Sialis lutaria*. *J. exp. Biol.*, **32**, 84–94.

Stein, F. (1838). Entwicklungs-Geschichte mehrerer Insectengattungen aus der Ordnung der Neuropteren. *Arch. Naturgesch.*, **4**, 315–333.

Stelzl, von M. (1991). Untersuchungen zu Nahrungsspektren mitteleuropäischer Neuropteren-Imagines (Neuropteroidea, Insecta). *J. Appl. Ent.*, **111**, 469–477.

Stelzl, M. (1992). Comparative studies on mouthparts and feeding habits of adult Raphidioptera and Neuroptera (Insecta: Neuropteroidea). *Proc. Fourth Internat. Symp. Neuropterology*, 341–347.

Stroot, P. (1986). Révision des Mégaloptères de la collection belge de l'I.R.S.N.B.- *Sialis nigripes* Ed. Pictet, 1865, Belg. n. sp. (Megaloptera, Sialidae). *Bull. Annls Soc. r. belge Ent.*, **122**, 195–201.

Sutcliffe, D. W. (1962). The composition of the haemolymph in aquatic insects. *J. exp. Biol.*, **39**, 325–343.

Sutcliffe, D. W. (1963). The chemical composition of haemolymph in insects and some other arthropods in relation to their phylogeny. *Comp. Biochem. Physiol.*, **9**, 121–135.

Syms, E. E. (1935). Biological notes on British Megaloptera. *Proc. S. Lond. ent. nat. Hist. Soc.*, 1934–5, 121–4.

Tjeder, B. (1944). A note on the food of the adult *Sisyra fuscata* F. (Neuroptera, Sisyridae). *Ent. Tidskr.*, **65**, 203–204.

Tjeder, B. (1968). A subfossil find of *Sialis* in Sweden. *Opusc. ent.*, **33**, 189–190.

Townsend, C. R. & Hildrew, A. G. (1979). Resource partitioning by two freshwater invertebrate predators with contrasting foraging strategies. *J. Anim. Ecol.*, **48**, 909–920.

Vaňhara, J. (1970). The taxonomy and faunistics of the Czechoslovakian species of the order Megaloptera. *Acta ent. bohemoslov.*, **67**, 133–141.

Vshivkova, T. S. (1985). Sialidae (Megaloptera) of Europe and the Caucasus. *Entomol. Obozr.*, **64**, 146–157. (in Russian).

Ward, P. H. (1965). A contribution to the knowledge of the biology of *Osmylus fulvicephalus*. *Entomologist's Gaz.*, **16**, 175–182.

Weissmair, W. (1994). Eidonomie und Ökologie zweier europäischer Schwammfliegen-Arten (Neuroptera: Sisyridae). *Entomol. Gener.*, **18**, 261–272.

Weissmair, W. & Waringer, J. (1994). Identification of the larvae and pupae of *Sisyra fuscata* (Fabricius, 1793) and *Sisyra terminalis* Curtis, 1854 (Insecta: Plannipennia: Sisyridae), based on Austrian material. *Aquatic Insects,* **16**, 147–155.

Wesenberg-Lund, C. (1917). *Furesøstudier. En Bathymetrisk Botanisk Zoologisk Undersøgelse af Mølleaaens Søer.* Kobenhavn. A. F. Host. 208 pp.

Withycombe, C. L. (1923). Notes on the biology of some British Neuroptera (Plannipennia). *Trans. R. ent. Soc. Lond.*, **1922**, 501–594.

Withycombe, C. L. (1925). Some aspects of the biology and morphology of the Neuroptera. With special reference to the immature stages and their possible phylogenetic significance. *Trans. R. ent. Soc. London.*, **1924**, 303–411.

Wootton, R. J. (1988). The historical ecology of aquatic insects: an overview. *Palaeogeography, Palaeoclimatology, Palaeoecology*, **62**, 477–492.

INDEX

Species names in parentheses are synonyms.

PUBLICATIONS OF THE FRESHWATER BIOLOGICAL ASSOCIATION

The publications listed on the following pages are some of those which are currently available for purchase. Order forms and further information may be obtained from:- **Dept. DWS, Freshwater Biological Association, The Ferry House, Far Sawrey, Ambleside, Cumbria LA22 0LP, UK.** The FBA regrets that it is unable to send books on approval.

SCIENTIFIC PUBLICATIONS (SPs)

SP 5. A KEY TO THE BRITISH SPECIES OF FRESHWATER CLADOCERA, WITH NOTES ON THEIR ECOLOGY, by D.J. Scourfield & J.P. Harding.
Third edition, 1966. (Reprinted 1994). Pp. 1-61. ISBN 0 900386 01 0.

SP 13. A KEY TO THE BRITISH FRESH- AND BRACKISH-WATER GASTROPODS, WITH NOTES ON THEIR ECOLOGY, by T.T. Macan.
Fourth edition, 1977. (Reprinted 1994). Pp. 1-46. ISBN 0 900386 30 4.

SP 17. A KEY TO THE ADULTS AND NYMPHS OF THE BRITISH STONEFLIES (PLECOPTERA), WITH NOTES ON THEIR ECOLOGY AND DISTRIBUTION, by H.B.N. Hynes.Third edition, 1977. (Reprinted 1993). Pp. 1-92. ISBN 0 900386 28 2.

SP 18. A KEY TO THE BRITISH FRESHWATER CYCLOPID AND CALANOID COPEPODS, WITH ECOLOGICAL NOTES, by J.P. Harding & W.A. Smith.
Second edition, 1974. Pp. 1-56. ISBN 0 900386 20 7.

SP 23. A KEY TO THE BRITISH SPECIES OF FRESHWATER TRICLADS (TURBELLARIA, PALUDICOLA), by T.B. Reynoldson.
Second edition, 1978. Pp. 1-32 + 1 colour plate. ISBN 0 900386 34 7.

SP 25. SOME METHODS FOR THE STATISTICAL ANALYSIS OF SAMPLES OF BENTHIC INVERTEBRATES, by J.M. Elliott.
Second edition, 1977. (Reprinted 1993). Pp. 1-160. ISBN 0 900386 29 0.

SP 27. A KEY TO THE FRESHWATER FISHES OF THE BRITISH ISLES, WITH NOTES ON THEIR DISTRIBUTION AND ECOLOGY, by P.S. Maitland.
1972. Pp. 1-139. ISBN 0 900386 18 5.

SP 29. TURBULENCE IN LAKES AND RIVERS, by I.R. Smith.
1975. Pp. 1-79. ISBN 0 900386 21 5.

SP 30. AN ILLUSTRATED GUIDE TO AQUATIC AND WATER-BORNE HYPHOMYCETES (FUNGI IMPERFECTI), WITH NOTES ON THEIR BIOLOGY, by C.T. Ingold. 1975. Pp. 1-96. ISBN 0 900386 22 3.

SP 31. A KEY TO THE LARVAE, PUPAE AND ADULTS OF THE BRITISH DIXIDAE (DIPTERA), THE MENISCUS MIDGES, by R.H.L. Disney.
1975. Pp. 1-78. ISBN 0 900386 23 1.

SP 33. DEPTH CHARTS OF THE CUMBRIAN LAKES, by A.E. Ramsbottom. 1976. Pp. 1-39. ISBN 0 900386 25 8. [Data are given for 15 lakes.]

SP 34. AN ILLUSTRATED KEY TO FRESHWATER AND SOIL AMOEBAE, WITH NOTES ON CULTIVATION AND ECOLOGY, by F.C. Page. 1976. Pp. 1-155. ISBN 0 900386 26 6.

SP 36. WATER ANALYSIS: SOME REVISED METHODS FOR LIMNOLOGISTS, by F.J.H. Mackereth, J. Heron & J.F. Talling. Second impression, 1989. Pp. 1-120. ISBN 0 900386 31 2.

SP 38. A KEY TO THE FRESHWATER PLANKTONIC AND SEMI-PLANKTONIC ROTIFERA OF THE BRITISH ISLES, by R.M. Pontin. 1978. Pp. 1-178. ISBN 0 900386 33 9.

SP 39. A GUIDE TO METHODS FOR ESTIMATING MICROBIAL NUMBERS AND BIOMASS IN FRESH WATER, by J.G. Jones. 1979. Pp. 1-112. ISBN 0 900386 37 1.

SP 41. A KEY TO THE BRITISH AND EUROPEAN FRESHWATER BRYOZOANS, by S.P. Mundy. 1980. Pp. 1-31. ISBN 0 900386 39 8.

SP 42. A KEY TO THE COMMONER DESMIDS OF THE ENGLISH LAKE DISTRICT, by E.M. Lind & A.J. Brook. 1980. Pp. 1-123. ISBN 0 900386 40 1.

SP 44. A GUIDE TO THE MORPHOLOGY OF THE DIATOM FRUSTULE, WITH A KEY TO THE BRITISH FRESHWATER GENERA, by H.G. Barber & E.Y. Haworth. 1981. (Reprinted 1994). Pp. 1-112. ISBN 0 900386 42 8.

SP 45. A KEY TO THE LARVAE OF THE BRITISH ORTHOCLADIINAE (CHIRONOMIDAE), by P.S. Cranston. 1982. Pp. 1-152 + 1 plate. ISBN 0 900386 43 6.

SP 46. THE PARASITIC COPEPODA AND BRANCHIURA OF BRITISH FRESHWATER FISHES: A HANDBOOK AND KEY, by G. Fryer. 1982. Pp. 1-87. ISBN 0 900386 44 4.

SP 47. A KEY TO THE ADULTS OF THE BRITISH EPHEMEROPTERA, WITH NOTES ON THEIR ECOLOGY, by J.M. Elliott & U.H. Humpesch. 1983. Pp. 1-101 + 1 plate. ISBN 0 900386 45 2.

SP 48. KEYS TO THE ADULTS, MALE HYPOPYGIA, FOURTH-INSTAR LARVAE AND PUPAE OF THE BRITISH MOSQUITOES (CULICIDAE), WITH NOTES ON THEIR ECOLOGY AND MEDICAL IMPORTANCE, by P.S. Cranston, C.D. Ramsdale, K.R. Snow & G.B. White. 1987. Pp. 1-152. ISBN 0 900386 46 0.

SP 49. LARVAE OF THE BRITISH EPHEMEROPTERA: A KEY WITH ECOLOGICAL NOTES, by J.M. Elliott, U.H. Humpesch & T.T. Macan. 1988. Pp. 1-145. ISBN 0 900386 47 9.

SP 50. ADULTS OF THE BRITISH AQUATIC HEMIPTERA HETEROPTERA: A KEY WITH ECOLOGICAL NOTES, by A.A. Savage.
1989. Pp. 1-173. ISBN 0 900386 48 7.

SP 51. A KEY TO THE CASE-BEARING CADDIS LARVAE OF BRITAIN AND IRELAND, by I.D. Wallace, B. Wallace & G.N. Philipson.
1990. Pp. 1-237. ISBN 0 900386 49 5.

SP 52. BRITISH FRESHWATER CRUSTACEA MALACOSTRACA: A KEY WITH ECOLOGICAL NOTES, by T. Gledhill, D.W. Sutcliffe & W.D. Williams.
1993. Pp. 1-176. ISBN 0 900386 53 3.

SP 53. A REVISED KEY TO THE CASELESS CADDIS LARVAE OF THE BRITISH ISLES, WITH NOTES ON THEIR ECOLOGY, by J.M. Edington & A.G. Hildrew.
1995. Pp. 1-134. ISBN 0 900386 55 X.

SPECIAL PUBLICATIONS (SPECs)

SPEC 1. A NATURAL HISTORY OF THE LAKES, TARNS AND STREAMS OF THE ENGLISH LAKE DISTRICT, by G. Fryer, 1991. ISBN 0 900386 50 9.
Pp. 384, landscape 184 x 248 mm, sewn and bound between soft covers.
[Recommended for beginners as an introduction to freshwater organisms.]

SPEC 3. EUTROPHICATION: RESEARCH AND APPLICATION TO WATER SUPPLY, edited by D.W. Sutcliffe & J.G. Jones, 1992. ISBN 0 900386 52 5.
Pp. 224, 186 x 245 mm, sewn and bound between soft covers.

SPEC 4. WATER QUALITY AND STRESS INDICATORS IN MARINE AND FRESHWATER ECOSYSTEMS: LINKING LEVELS OF ORGANISATION (INDIVIDUALS), POPULATIONS, COMMUNITIES), edited by D.W. Sutcliffe, 1994. ISBN 0 900386 54 1. Pp. 182, 186 x 245 mm, sewn and bound between soft covers.